MINING IN A MEDIEVAL LANDSCAPE

This book examines for the first time how a local industry revolutionized the way mining was carried out in medieval Britain and beyond.

Mining in a Medieval Landscape explores the history and archaeology of the late-medieval royal silver mines at Bere Ferrers in the Tamar valley in Devon. It compares their impact on the landscape with less intensive, traditional mining industries. The analysis of maps and documents together with archaeological field-survey work allows the mining landscape to be reconstructed in remarkable detail, including Bere Alston – probably the first purpose-built mining town in Britain.

Stephen Rippon is Professor of Landscape Archaeology at the University of Exeter. He is a Fellow of the Society of Antiquaries of London and a Member of the Institute of Field Archaeologists. He makes regular television appearances. **Peter Claughton** has worked as an economic historian since gaining his doctorate, becoming an expert in his field. He is Conservation Officer of the National Association of Mining History Organisations and an active member of the international industrial heritage committee TICCIH. **Chris Smart** was responsible for field survey and historic landscape analysis on the Bere Ferrers project. He discovered, and subsequently directed excavations of, the Roman fort at Calstock. He is now a Project Archaeologist with Exeter Archaeology.

Frontispiece: Aerial photograph of Bere Ferrers peninsula including the mine workings, Bere Alston and the high ground beyond (reproduced courtesty of Barry Gamble, The Tamar Valley AONB).

Mining in a Medieval Landscape

The Royal Silver Mines of the Tamar Valley

STEPHEN RIPPON,
PETER CLAUGHTON
AND CHRIS SMART

UNIVERSITY
of
EXETER
PRESS

First published in 2009 by
University of Exeter Press
Reed Hall, Streatham Drive
Exeter EX4 4QR
UK
www.exeterpress.co.uk

© 2009 Stephen Rippon, Peter Claughton and Chris Smart

The right of Stephen Rippon, Peter Claughton and Chris Smart to be identified as authors of this work has been asserted by them in accordance with the Copyright, Designs and Patents Acts 1988.

British Library Cataloguing in Publication Data
A catalogue record for this book is available from the British Library.

ISBN 978-0-85989-828-7

Typeset in Stempel Garamond, 10.5 on 14 by
Carnegie Book Production, Lancaster

Contents

Illustrations	vi
Acknowledgements	ix
Abbreviations	xi
Glossary	xii
1 Introduction: the impact of industry on the medieval landscape	1
2 Earth's riches: metal resources in medieval Devon	13
3 Silver production in medieval England and the Devon mines	45
4 The extraction and processing of silver-bearing ores	71
5 Fuelling the industry: the management of water and woodland	101
6 The mining community and its impact on the wider landscape	121
7 Discussion and conclusions	160
Notes	165
Sources and bibliography	183
Index	205

Illustrations

Figures

Frontispiece: Aerial photograph of Bere Ferrers peninsula		ii
1.1	The nineteenth-century Carn Galver tin mine north of Rosemergy hamlet, Cornwall	2
1.2	Location of Bere Ferrers in the Tamar valley south-west of Dartmoor	5
1.3	The medieval earthworks on Lockridge Hill	10
2.1	The major mineral resources of Devon and adjacent areas	14
2.2	Interpretation of the geophysical surveys of workings on Harris's Lode, Combe Martin	15
2.3	The Bampfylde copper mine in North Molton	25
2.4	The tin streamworks at Lydford, south-west of Dartmoor	31
2.5	The medieval iron industry of the Blackdown Hills	38
2.6	The iron, copper, and lead/silver industries of the Exmoor region	41
3.1	Silver penny of Henry II's reign struck at the Carlisle mint in the 1160s	47
3.2	Silver penny of Edward I struck at the London mint in the late 1290s	47
3.3	Continental silver mines	51
3.4	Employment structure of the Devon silver mines in the fourteenth century	59
3.5	Medieval woodland, mining and processing sites in the Bere Ferrers region	61
3.6	The mining landscape of Combe Martin, North Devon	64

3.7	Named sections of the Bere Ferrers mines and to whom they were farmed out in 1451–53	68
4.1	The value of recorded silver and lead production from the Devon mines 1292–1344	72
4.2	The mine workings in Cleave Wood	76
4.3	Documented mine workings in Bere Ferrers	78
4.4	Location of the cross-cutting adits at Bere Ferrers	80
4.5	Mine workings and adit at Furzehill shown on the 1737 estate map	81
4.6	Surveyed earthworks of the air shaft at Furzehill	82
4.7	Late medieval silver mines and trial workings in South-West England	85
4.8	The breaking up of ore-bearing rock	88
4.9	Smelting techniques at Bere Ferrers in the late medieval period	89
4.10	Aerial view of Calstock	95
4.11	Results of the magnetometer survey south of Calstock parish church	96
4.12	The equipment required for the refining of lead to produce silver	98
5.1	Maristow and Blaxton Wood from across the Tavy estuary	104
5.2	The construction and assembly of a suction-lift pump	111
5.3	The route of the Lumburn leat	112
5.4	The line of the Lumburn Leat at Hocklake Farm	114
5.5	The Lumburn Leat at Broadwell	115
5.6	The Lumburn Leat in Shillamill Wood	116
5.7	A deep cutting for the Lumburn Leat at Raven's Rock	116
5.8	The leat tunnel below Raven Rock	116
5.9	Plan of the cuttings at Raven's Rock	117
5.10	The Lumburn Leat south of Raven's Rock	118
6.1	Historic landscape characterisation of Bere Ferrers and adjacent areas	122
6.2	Vertical aerial photograph of Bere Alston	128

6.3 Bere Alston on the first edition of the Ordnance Survey Six Inch maps — 129

6.4 Bere Alston on the 1737 estate map — 130

6.5 Bere Ferrers parish: land ownership as recorded in the Tithe survey — 131

6.6 Bere Ferrers parish: land holding as recorded in the Tithe survey — 132

6.7 Bere Ferrers parish: land occupancy as recorded in the Tithe survey — 133

6.8 Characterisation of the settlement pattern in Bere Ferrers and adjacent parishes — 135

6.9 The former medieval park in the far south of Bere Ferrers parish — 137

6.10 Extract from the 1737 estate map showing the small hamlet at Bere Ferrers — 139

6.11 The planned medieval borough at South Zeal — 142

6.12 Bere Alston on the Tithe map of 1845 — 143

6.13 Nineteenth-century settlements in Bere Ferrers parish — 146

6.14 Metherell and St Dominick showing the extensive former open fields — 148

6.15 Whitsam on the First Edition Ordnance Survey Six Inch map — 150

6.16 Whitsam on the 1737 estate maps — 151

6.17 Places in Bere Ferrers parish referred to in the late fifteenth-century 'confessor's itinerary' — 152

6.18 The relative size of settlements in Bere Ferrers parish based on the late fifteenth-century 'confessor's itinerary' — 154

Tables

3.1 Account of William de Wymondham 20–25 Edw. I [1292–97] — 55–59

4.1 Extracts from the wage roll for the mines at Bere Ferrers in 1342–43 — 73–74

4.2 Bole smelting costs in 1342–44 — 91–92

6.1 Defining features of the historic landscape character types — 124–27

Acknowledgements

THIS BOOK is the outcome of a major research project that lasted from May 2006 through to March 2008. It was generously funded by the Leverhulme Trust with additional support from the University of Exeter. The authors would like to take this opportunity to thank the many individuals, landowners, and organisations who helped during the course of the project: Mrs Betsy Gallup, Jerry Venning, Graham Tew, The Woodland Trust, Mrs Rees of Pengarth, Paul Wiseman (agent to the Mount Edgcumbe Estate), Mrs Sargent, Mr and Mrs Cole, Trevor Jeans and Tony Viggers, Mr and Mrs Neave, Hugh Harrison, Malcolm Collingridge, Mark Snellgrove of Tavistock Woodlands, Mr and Mrs Maurice Gerry, Peter Baker, Mr and Mrs Edwards of Colcharton Farm, Sandra Anstey, Mr Tait of Millhill Quarry, Joe Hess of the Maristow Estate, Mr Turpin of Pound Farm, Dr Charlie Moon, Mrs G.M. Astruc, John Start and Chris Facey, Trevor Dunkerley for permission to use geophysical survey results from Combe Martin, Mick Warburton and Roger Burton for assistance at Combe Martin, Alasdair Neill, The Calstock Parish Archive, Peter Mayer and Richard Bass for their help with early work on the medieval documents, the staff of the county record offices in Exeter and Truro, particularly Paul Brough and Deborah Tritton for a digital image of ME2424, and the National Archives at Kew, along with Janis Heward and Helen Rance who contributed their time to the detailed survey work. We would also like to thank Steve Hartgroves of Cornwall County Council for granting permission for us to reproduce Figure 4.10, Stephen Henley of Resources Computing International Ltd for permission to use the images in Figures 4.12 and 5.2, and Martin Allen of the Fitzwilliam Museum, Cambridge, for providing the coin images in Figures 3.1 and 3.2.

Towards the end of the project, geophysical survey at Calstock revealed what appeared to be part of a Roman fort, and in January 2008 a small excavation was carried out to test this hypothesis. We are once again grateful to the Leverhulme Trust and the University of Exeter for supporting this work. The excavations were directed by Chris Smart, and carried out with Peter Claughton and a team of skilled volunteers – Helen Rance, Simon Hughes, Alex Farnell, Catherine Rackham, Janis Heward, Naomi Payne, Colin Wakeham and Graham Tait – whom we would like to thank for their valuable contribution.

Abbreviations

BGS	British Geological Survey
BL	British Library
Cal. Charter R.	*Calendar of Charter Rolls*
Cal. Close R.	*Calendar of the Close Rolls*
Cal. Fine R.	*Calendar of the Fine Rolls*
Cal. Inq. Misc	*Calendar of the Inquisitions Miscellaneous*
Cal. Inq. Post Mortem	*Calendar of the Inquisitions Post Mortem*
Cal. Liberate R.	*Calendar of the Liberate Rolls*
Cal. Pat R.	*Calendar of the Patent Rolls*
CRO	Cornwall Record Office
DCC HER	Devon County Council Historic Environment Record
dGPS	differential Global Positioning System
DRO	Devon Record Office
GIS	Geographical Information System
Letters & Papers Hen. VIII	*Letters and Papers, Foreign and Domestic, Henry VIII*
TNA: PRO	The National Archives: Public Record Office
WRO	Wiltshire Record Office
WYAS	West Yorkshire Archive Service

Glossary

Adit 'Avidod' in the medieval accounts for the Devon mines and 'Audit' on the eighteenth-century mapping of Bere Ferrers – a virtually horizontal tunnel usually driven to effect free drainage of water to surface.

Assart an enclosure within woodland, converted to arable.

Blackwork Smelting residues, or slag, which still contained some metals and required re-smelting.

Bole A smelting process relying on the wind to provide the draught required to achieve a high enough temperate for separation of lead from its ore. Used in the Devon mines until at least the 1340s and commonly used in non-agrentiferous lead mining fields until the mid-sixteenth century.

Burgage appertaining to a borough.

Burgess a citizen of a borough holding full municipal rights.

Cerussite Lead carbonate, referred to as 'white ore' in the *Exchequer Accounts*.

Cupellation The method of refining silver from lead metal by oxidising the lead to form litharge (lead oxide) but leaving the silver as a metal.

Deadwork The development work required to open up and/or drain the ore-bearing deposits. Deadwork was, by definition, unproductive.

Development work see 'deadwork' (above).

Dish An undefined measure of volume, used for ore in the medieval Devon silver mines, and commonly used in other lead and silver mining fields. In the north Pennine mines of the twelfth / thirteenth century a dish 'should contain as much ore as a man can lift from the ground' (*Calendar of Documents relating to Scotland*, vol. 3 p. 295).

Dressing The process of preparing ore, as mined, for smelting by separating it from the waste or 'gangue' material.

Fathom A linear measure used in mining, being six feet or 1.8 metres.

Glossary

Foot   A measure of weight used for lead metal in the Devon silver mines, being 70 lbs mercantile or 30.63 kg.

Galena   Lead sulphide, referred to as 'black ore' in the *Exchequer Accounts*.

Gangue   This is the waste mineral which surrounds or is mixed with the metal-bearing ore. It has to be separated from the metalliferous ores, either by hand or using gravity separation in water, before the latter can be smelted.

Hutt   A smelting process used in the Devon mines in the 1290s, it used a formed of liquation (based on the affinity of silver to lead) to extract silver from difficult ores. The term means 'smelting house' in German and could indicate the origins of the process.

Litharge   a protoxide of lead formed by exposing metallic lead to a blast of air.

Load   A lead ore measure made up of nine dishes.

Pre-emption   The right of a mineral owner or, in some cases, the Crown to take the produce of the mines at a set price.

Refining   Separating the silver from lead – see cupellation (above).

Saiger process   a multi-stage smelting technique for argentiferous (silver bearing) copper ores in which metallic lead was introduced, relying on the affinity of silver for lead to de-silver the copper.

Smelting   Bulk extraction of metals from their ores using heat.

Turnbole   a wind-blown 'bole', lead smelting hearth, mounted on a moveable timber platform, allowing it to be turned to take advantages of changes in wind direction.

Chapter 1

Introduction
The impact of industry on the medieval landscape

THE LANDSCAPE we live in – whether it be urban or rural – forms a crucial part of our sense of place and identity. We are rightly proud of the role that Britain played in the industrial revolution, but equally cherish our open countryside. Indeed, there is a common perception on the part of the non-specialist that these two aspects of our landscape – the industrial and the rural – are quite separate, with industry belonging very much to the townscape. Many would also draw a contrast between ever-changing urban landscapes, with their hectic lifestyles and constant regeneration, and the slower pace of life within a peaceful and relatively unchanging rural countryside. In places, this may have been the case, and the pattern of fields, roads, and settlements across most of the lowlands in the south-west of England do indeed date back to before the Norman Conquest, forming an apparently peaceful and unchanging countryside (Figure 1.1). In the far west of Cornwall – the area known as West Penwith – the landscape is even older, having its origins in the late prehistoric period, and this now rural region could not be more different in character to the nearby bustling conurbation of Camborne and Redruth. This was the centre of the Cornish copper and tin industry for many centuries and, although fallen on hard times in recent years, could yet undergo a revival in its fortunes if mining were to resume as global prices for tin rise.

Appearances can, however, be deceptive, and in many regions research into the history of our countryside is revealing the extent to which change has been a major theme – even in rural areas. The South West is a fine example, for within the predominantly rural landscape there

Figure 1.1: The familiar image of mining in south-west England. The nineteenth-century Carn Galver tin mine north of Rosemergy hamlet, in Morvah parish, Cornwall, symbolic of what was once a dynamic industrial landscape. The strongly lynchetted fields appear to have been in use since the late prehistoric period and suggest slower pace of change in the countryside.

PHOTOGRAPH: STEPHEN RIPPON

are occasional reminders of what was once a more strongly industrial past: the chimneys and engine houses that lie above long-abandoned mines but which still characterise the landscape in some upland and coastal districts (Figure 1.1). One example of a now peaceful rural landscape that has a hidden industrial past is the Bere Ferrers peninsula, between the Tamar and Tavy valleys on the Devon–Cornwall border. Despite its close proximity to Plymouth, today this is an area of quiet rural countryside with a scatter of isolated farmsteads and just a single small town at Bere Alston. It is difficult to believe that this was once the centre of England's medieval silver mining industry, but using the exceptionally rich documentary sources that have survived, and relating these documents to physical evidence on the ground, the story of this nationally important industry, and the impact that it had on the local area, will be explored for the first time.

The impact of mining on the landscape

In the post-medieval period mining was to develop primarily as a large-scale activity with, in some cases, associated urbanisation. This was particularly the case in the coal industry where the opening up of concealed coalfields in the second half of the nineteenth century demanded the provision of dedicated housing in support of its long-term development. The majority of non-coal mining sites, notably the non-ferrous metal mines, were rural in location, but from the mid-eighteenth century onwards the smelting of the ores was increasingly carried out away from the mines and close to the coalfields which were the source of fuel or close to suitable transport nodes. The result of this was that industry became a largely urban activity concentrated into a relatively small number of locations.

In the medieval period, in contrast, the scale of industrial production, and the demands that it placed upon the landscape were on an altogether different scale: the more limited demand for labour meant that there was no requirement for major settlements, and as woodland (producing charcoal) and water (for milling) were the major sources of fuel, it was the countryside that accommodated most industry. Historians have suggested that the medieval mining industry – both the extraction of metal ores from the ground and the processing and smelting of these ores in order to produce metal – was carried out by communities who were engaged in both farming and mining (a system known as dual-occupancy). Various suggestions have been made as to how this might have worked including mining being a small-scale operation carried out on a seasonal basis in between the demands of ploughing and lambing in the spring and harvesting in the late summer. Even those historians who suggest that miners would not have been able to produce all their own food see them as living in essentially agricultural communities (this is discussed further in Chapter 2).

These models of part-time mining would have had relatively little impact on the wider landscape. There is, however, one particular industry that stands out as potentially different. In the increasingly commercialised economy of twelfth- and thirteenth-century England, the Crown needed silver in order to mint coins (discussed further in Chapter 3). Silver-bearing ores were found in various parts of the kingdom, including the Mendip Hills in Somerset, around which there was a particular concentration of late Anglo-Saxon mints. By the twelfth century there was significant production from the north Pennines, but by the early thirteenth century it appears that little or no silver was being mined in England and it was continental European silver, drawn into England through an expanding export trade, which provided for the significant increase in the volume of coin in circulation. This reliance on continental sources stimulated the search for new deposits in England, and prospecting on behalf of the Crown in the mid-thirteenth century identified, but failed to realise production from, silver-bearing deposits in Devon. In 1292, however, the Crown opened up new mines (*Minera regis de Birlaund et de Combe martin*) in an otherwise obscure Devon manor of 'Birland', now known as Bere Ferrers in the south of the county, and at Combe Martin in the north.

The opening of these mines marked a significant change in the way that mining was carried out in England. In the past, mineral resources had been exploited by local communities whose work was controlled through customary practices, whereas in these new Devon enterprises the Crown chose to operate in a different way, by directly managing the mines and employing its own miners who were overseen by salaried royal officials. Although the Combe Martin enterprise closed in 1296, the Bere Ferrers mines remained under direct royal control until at least 1349, whereafter they were leased out until they closed in the mid-sixteenth century. It was not, however, just the way in which the Bere Ferrers mines were managed that singled them out as being different: it was also the scale of the operation. There was no tradition of deep mining in Devon, as its most famous industry at that time – tin – was based on working surface deposits. The silver-bearing ores at Bere Ferrers, in contrast, had to be mined underground and so men experienced in hard-rock mining had to be brought in from areas such as the Peak District and north-east Wales. In 1298, for example, over three hundred miners were employed at Bere Ferrers, with a further hundred men engaged in drainage works. There was also a distinct difference in the spatial context within which this industry developed. Until the development of the Bere Ferrers mines, most mineral resources exploited in the medieval period occurred extensively across relatively wide areas: four hundred men employed across the surface tin deposits of Devon and Cornwall could easily have been accommodated within what were otherwise landscapes of agriculture and upland grazing. What was different at Bere Ferrers was the strongly focused nature of the mining operation: the silver-bearing deposits were located along a discrete section of a single lode where the richest mineralisation outcrops at surface in a line just two kilometres long. This will have led to a far more concentrated mining operation than had previously been the case elsewhere, and it is the impact that this had on the landscape which will be a major theme of this study.

The landscape of Bere Ferrers

The parish of Bere Ferrers is almost wholly defined by the rivers Tamar, on its western side, and Tavy on its eastern side, with the former marking the boundary between the counties of Devon and Cornwall. It lies some ten kilometres to the south-west of Tavistock, on the western fringes of Dartmoor, and nine kilometres north of Plymouth Sound and the English Channel. These two rivers define the peninsula of Bere Ferrers, which is seven kilometres long and up to five kilometres wide (Figure 1.2). The peninsula is topographically isolated, being connected with the remainder of Devon by a narrow neck of high ground, rising to 180 m at Morwell Down, and which is at most one and a half kilometres wide. The peninsula comprises a gently undulating plateau rising from around 80–100 m OD in the west to 150 m OD in the east. The eastern side of the peninsula is marked by steep slopes that fall away to the river Tavy, which today

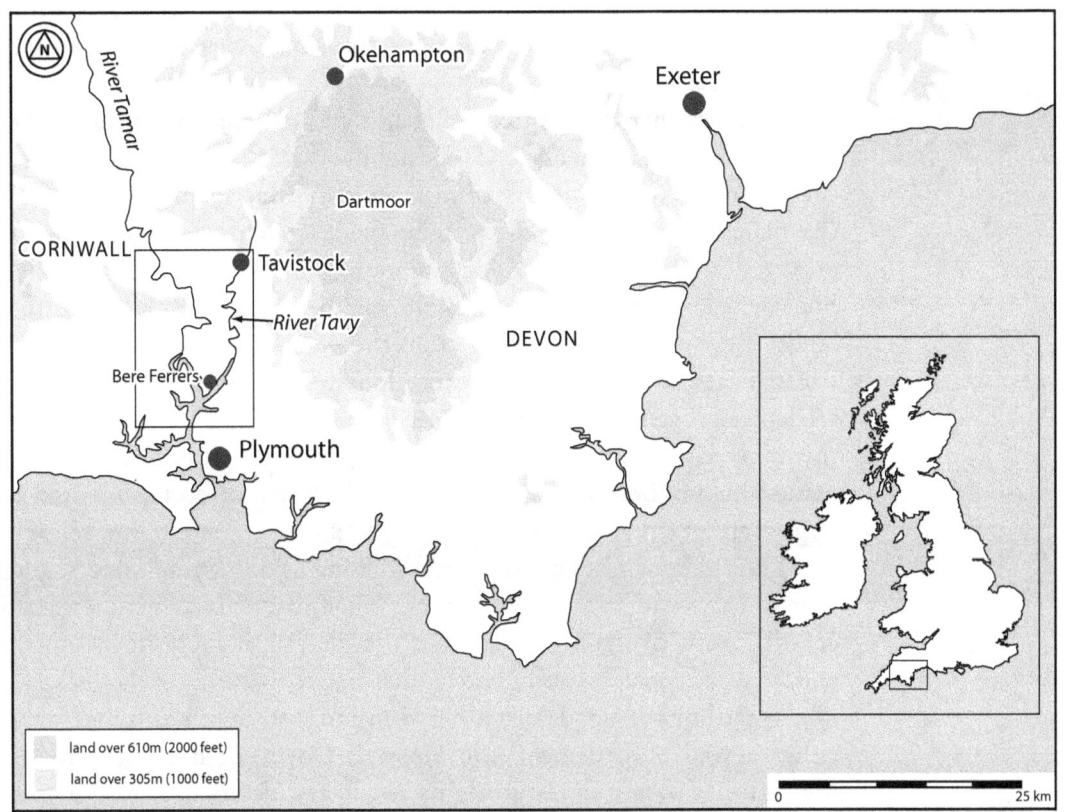

Figure 1.2: Location of Bere Ferrers in the Tamar valley south-west of Dartmoor.

are heavily wooded, as they have been since at least the early eighteenth century when the parish was first mapped. Two steep-sided narrow valleys, Liphill Lake[1] in the west and Hallowell Lake in the east, divide the southern part of the peninsula into three. The manor house and adjacent parish church of Bere Ferrers are located on the banks of the river Tavy in the far south of the peninsula, next to an area of lower, gently undulating land that was once occupied by a medieval deer park (see Chapter 6).

The peninsula is underlain by slaty mudstones, shales, and sandstones of the Upper Devonian and Lower Carboniferous Tavy Formation, with soils belonging to the Denbigh 1 Association (well-drained, fine loamy and fine silty soils with some slight seasonal waterlogging).[2] Silver-bearing lead ores, associated with fluorite, occur within a number of north–south oriented mineralised cross-courses or lodes with a steep easterly dip, which are cut and displaced by south dipping fracture zones.[3] Only one of those lodes, often referred to as the 'eastern cross-course' and running through the western part of the peninsula where the rich mineralisation outcrops at surface for some two kilometres between Lockridge Hill in the north and Cleave, close to the river Tamar, in the south, was worked in the medieval period.

When first mapped in the eighteenth century, the character of the fields, roads, settlements and commons varied significantly across the peninsula (this is discussed in more detail in Chapter 6). Most of the plateau was covered in medium-sized, roughly rectilinear fields arranged around a series of long, sinuous, and often broadly parallel boundaries, some of which were followed by roads. Dotted across the landscape, but usually set back from the main roads, were a series of isolated farmsteads, with two exceptions to this otherwise dispersed settlement pattern being the hamlet of Bere Ferrers itself – which in the nineteenth century comprised the medieval parish church, the manor house (Bere Barton), and a small cluster of cottages – and a larger settlement at Bere Alston which included a church, inn, school, three non-conformist chapels, terraced housing along four streets, and a small cluster of farms and cottages. When mapped in the nineteenth century some of the higher areas of the plateau, notably in the north and east of Bere Alston, had a distinctive pattern of larger and more rectilinear fields laid out between long, straight roads and were either devoid of settlement (e.g. Morwell Down) or had place-names indicative of relatively late origins (such as 'Newhouse' east of the

Hallowell Lake). This was clearly an area of relatively late enclosure of what had been common pasture, in between the steep slopes of the Tavy and Hallowell Lake valleys.

The Bere Ferrers Project: an interdisciplinary approach to studying mining landscapes

This book outlines the results of a two-year study of the Bere Ferrers landscape, funded by the Leverhulme Trust and the University of Exeter, and carried out between 2006 and 2008.[4] It represents a collaboration between an economic historian (Peter Claughton) and two landscape archaeologists (Stephen Rippon and Chris Smart), and reflects the desire of all three authors to move beyond traditional studies of industrial archaeology and mining history towards a more integrated understanding of the landscape. Despite its economic and social importance, the study of early industry has not always received the attention that it deserves within archaeology and history, which in part reflects the wider fragmentation in scholarship that is all too common for the medieval period. Many specialist groups[5] have interests that focus on aspects of landscape and industry, although few have attempted to study both those subjects during the medieval period (two exceptions being the European Union funded Landmarks programme and the work of the Royal Commission on the Ancient and Historical Monuments of Wales).[6] Specialist groups are not in themselves a bad thing – it is always useful for those interested in a particular field of study to get together and discuss their common interests – but it becomes a problem if the results of this research struggle to permeate beyond the membership of that specialist group. One, albeit crude, way of assessing the impact of a particular discipline on the wider field of scholarship is to review its inclusion in major journals, and for the study of industry this makes somewhat depressing reading. In the sixteen volumes of *Landscapes*, published between 2000 and 2008, just two out of the ninety-five papers relate directly to industry. In the journals *Landscape History* and *Medieval Archaeology* the study of industry is also under-represented compared to work on agriculture, rural settlement, towns, and elite landscapes such as castles and parks. What is also striking is that journals that specialise in the study of industry, such as *Historical Metallurgy* and *Industrial Archaeology Review*, similarly have relatively little to say with regard to the impact that industry had on the wider landscape: the former is dominated by the study of technology and analysis of artefacts, while the latter is, by definition, concerned with the processes of the industrial revolution and more recent times.[7] Medieval historians generally steer clear of industry, but when the period is studied from the perspective of its economic history, industrial production does receive greater attention, although primarily in terms of urban organisation rather than its impact on rural society. Industrial archaeologists have largely concentrated on individual sites, but there is an increasing move to consider their relationship within the landscape, and that has been defined as an essential element of the discipline.[8] Newell and Walker, for example, have explored in depth the medieval origins of industry on Tameside, through its social structure – linked to lordship and landscape – developing the so-called 'Manchester Methodology', but that work is not widely known and its universal application has been questioned.[9] Mining has its fair share of coverage in both the academic and specialist historical journals, particularly for the post-medieval period, although few authors appear to consider themselves capable of addressing the medieval period or its landscape impact.

In this study of medieval mining, however, a different approach – more interdisciplinary and landscape-focused in its philosophy – has been adopted. Using Devon, a county rich in mineral resources, as a case-study, the aim has been to contrast the traditional small-scale mining activity of the medieval period with the unprecedented operations at Bere Ferrers. Devon is an area in which the mining of tin and, to a lesser extent, copper in the post-medieval period are well known, and the importance of those industries is reflected in the recent designation of the 'Cornwall and West Devon Mining Landscape' as a UNESCO World Heritage Site. In contrast, the exploitation of Devon's other metal resources such as gold, iron, and lead, has been largely neglected and so the early exploitation of all these metals is reviewed in Chapter 2 in order to characterise the background of traditional medieval mining in the region and the impact that it had on the landscape.

In common with the non-tin industries, the silver mines at Bere Ferrers, while touched upon in works of general mining history and industrial archaeology[10] had, until recently, been neglected by archaeologists and historians alike, even though the documentary sources are remarkably rich.[11] These archives include the Exchequer Accounts (E101) and other enrolled records of the state held in the National Archives, along with various transcripts in the British

Library and those held locally in the Calstock Parish archive.[12] An additional document that has proved particularly valuable is a list of settlements in Bere Ferrers parish now held in the Exeter Cathedral archives.[13] A variety of later sources also sheds light on the medieval landscape, including a survey of 1649–50 which lists the woodland at this period, and various other manorial and estate papers from the seventeenth and eighteenth centuries including an estate map of the manor of Bere Ferrers drawn up in 1737 for Lord Hobart. The Tithe map of 1845 provides the first comprehensive mapping of land holding and occupancy for the whole parish. Various archives relating to eighteenth- and nineteenth-century mining in the area have also been important in providing information on the extent of the medieval workings when they were abandoned.

The significance of the cartographic sources is not just that they show the location of some of the abandoned medieval mine workings, but that they contain large numbers of field- and place-names, many of which can be related to places referred to in the medieval documents. The mapping of these documented places allows the various components of the medieval mining landscape – the mines, processing sites, and associated infrastructure – to be mapped and related to surviving evidence on the ground, so giving a spatial and landscape dimension to the industry. A variety of physical evidence for the medieval landscape survives. Along the lode itself there are large numbers of earthworks, notably extraction shafts, air shafts for ventilation, and spoil heaps, some of which appear to be medieval in date (e.g. Figure 1.3). These can be distinguished from later mine workings dating to the late eighteenth to nineteenth centuries through their morphology (being smaller in scale), location (being positioned where the lode outcrops on the surface), their stratigraphic relationships to later working (being partly beneath later spoil), and absence of the boiler clinker associated with later post-medieval mining. The well-preserved nature of these earthworks – they are now mostly under woodland – and the way that they occur along a discrete alignment, means that they are, in effect, a single linear earthwork some two kilometres long. Another impressive linear earthwork is that of the Lumburn Leat which runs for sixteen kilometres from Ogbear, west of Tavistock, down the Lumburn and Tavy Valleys, around Morwell Down, and across to Lockridge Hill. The leat was constructed between 1470 and 1480 in order to power suction-lift pumps used to drain the mines, and in several places had to be tunnelled through bedrock, making it one of the most impressive yet previously unrecognised feats of medieval engineering in Devon. Finally, we must not forget that the modern patterns of fields, roads, and settlements are themselves mostly medieval in origin: documentary sources show that most of the farmsteads mapped in 1737 existed in the fourteenth and fifteenth centuries, while both the Lumburn Leat and the mine workings clearly cut through, and so post-date, the field

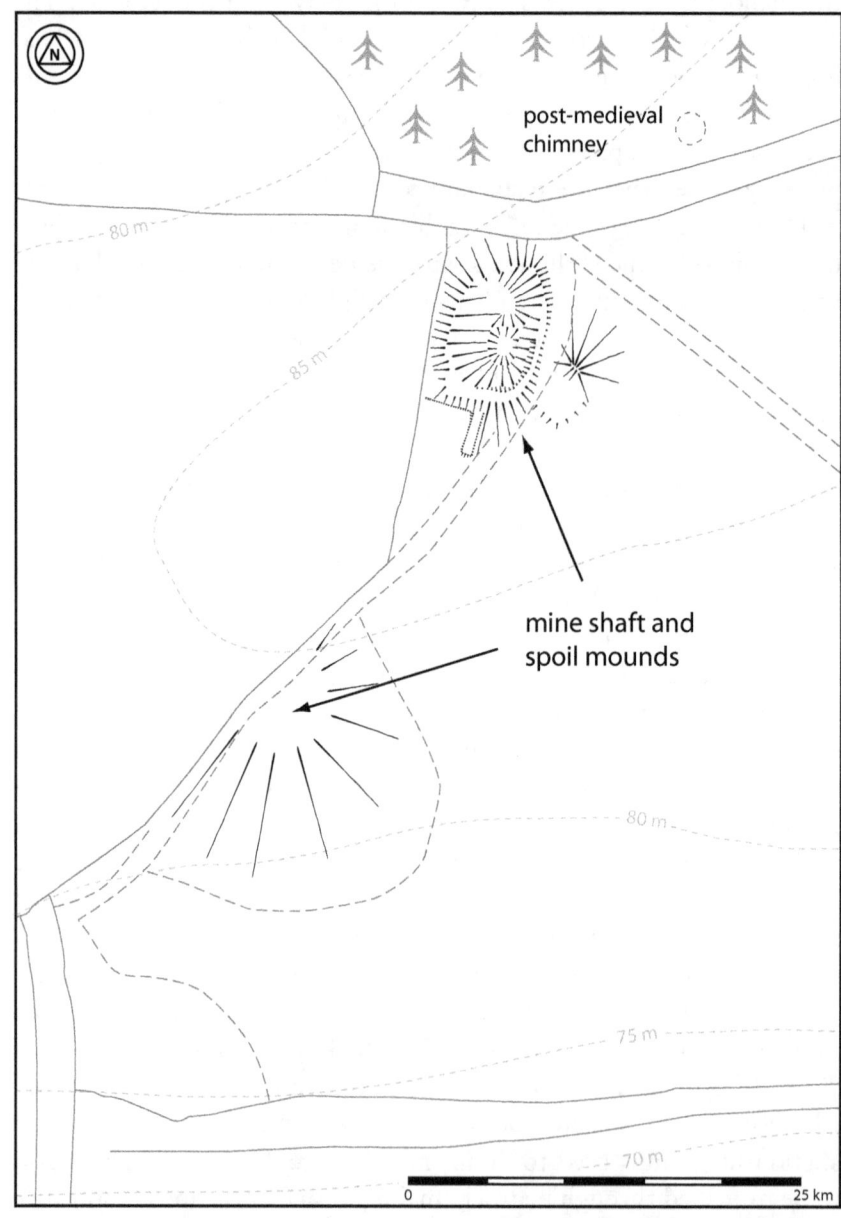

Figure 1.3: Earthworks on Lockridge Hill: the spoil from medieval shallow shaft working.

systems on this part of the peninsula, showing that they must have been laid out before the late medieval period. Taken altogether, this makes for a remarkably well-preserved medieval landscape.

Mining in a medieval landscape begins, in Chapter 2, with a review of the history of medieval mining in Devon. The contrast is drawn between those metals whose ownership rested with the landowner and which were worked subject to the regulations of local custom, and the more valuable metals (copper, lead, silver, and gold) over which the Crown exercised a right of prerogative. It will be shown that the former were worked on a relatively small scale, which made them accessible to part-time miners-cum-farmers, in contrast to the latter – and in particular silver – which was a larger-scale and capital intensive operation under Crown control. In Chapter 3, the mining of silver in medieval England is reviewed and placed in the wider context of its production across central and western Europe. The way in which the Bere Ferrers workings were managed is described, along with what is known about Devon's other silver mines at Combe Martin. Chapter 4 will describe the development of the silver mining landscape at Bere Ferrers including evidence for the extraction and processing of ores. The production of silver was carried out in essentially five stages. First there was the development work, referred to as 'deadwork' in the mine accounts. This included sinking shafts and cutting the levels required to open up the ore deposits and allow for drainage. The ore-bearing deposit itself was then mined and brought to the surface, where it was broken up and sorted (a process sometimes known as 'dressing') in order to remove the waste material (known as 'gangue'). This dressed ore was then smelted to produce a lead metal rich in silver which was then refined to recover the silver. Both the smelting and the refining required large amounts of fuel which meant that those activities were often carried out some distance from the mines, where suitable supplies of wood and charcoal could be obtained.

Chapters 5 and 6 will consider the impact that the mines had on the wider landscape both in terms of the demand for energy – supplied by wood and water – and support of the mining community. By the early years of the fourteenth century a major programme of capital expenditure was in place, taking the workings at Bere Ferrers well below the water-table with a consequent requirement for well-planned drainage. A continued demand for silver, particularly during the bullion crisis of the mid-fifteenth century, encouraged deeper working of the Bere Ferrers mines, and the attendant high costs of manual drainage stimulated the introduction of innovative mechanised pumping by 1480. The demand for structural timber in the mines, and as fuel for the smelting and refining processes, meant that woodland was exploited in a wide area around Bere Ferrers including for charcoal production. Water transport was used to supply the mines and smelting sites, which gave rise to expenditure on boat repairs, and a ropeworks was established to satisfy the requirements for haulage within the mines.

The mining community itself, and the wider landscape within which it existed, is discussed in Chapter 6. Medieval mining has traditionally been seen as a small-scale, part-time occupation, and so it might be assumed that the altogether different scale of operations seen at Bere Ferrers – and notably the large, predominantly full-time workforce – had a different impact upon the landscape. There is, however, surprisingly little evidence for this. The small market town at Bere Alston appears to have been established in order to support the mining industry, and miners were accommodated both there and within the existing rural settlement pattern rather than in specific mining villages.

Overall, it is hoped that this study will demonstrate the advantages of a new approach towards industrial archaeology and mining history through the integration of the two disciplines within a wider study of the landscape.

Chapter 2

Earth's riches
Metal resources in medieval Devon

IN THE MEDIEVAL PERIOD Devon, and its immediate borders, had perhaps the greatest range of metal resources of any region in England (see Figure 2.1). The tin deposits of Devon and Cornwall were unique as the premier source of that metal in Europe. Its lead/silver mines, while not as productive as the twelfth-century workings in the north Pennines, were England's only significant source of newly mined silver from the demise of the northern mines in the late twelfth century, until the sixteenth century. The region's copper ores, which were exploited in a small way from at least the fourteenth century, were to become the focus of a flourishing industry from the late seventeenth century and a world leader by the mid-nineteenth century. An iron industry, of more than local importance from the Roman occupation onwards, was sustained on ores mined on Exmoor and its borders, and in the Blackdown Hills. In some areas, metal production competed with wool and textiles as a major source of income with tin, in particular, contributing significantly to England's export earnings.[1] Some sectors of metal mining continued as important, expanding, elements within the economy of South-west England through to the nineteenth century, while others were at their peak in the medieval period; but all have left their mark on the Devon landscape.

Archaeological investigation relevant to early metal mining and processing in Devon has largely focused on tin and iron. Some research focusing on these two minerals has incidentally provided valuable information on the exploitation of other metals, but little detailed work has specifically been carried out on lead and silver, and virtually nothing on copper. Cornwall and West Devon's successful World Heritage bid has built on earlier work to provide an assessment of the mining archaeology in the Tamar valley and adjoining areas, but that was in support of a bid based on the modern, post-1750, expansion of the industry. Earlier mining is acknowledged but has not received the same detailed attention as later tin, copper, and arsenic production. Only ten years ago, when *The Archaeology of Mining and Metallurgy*

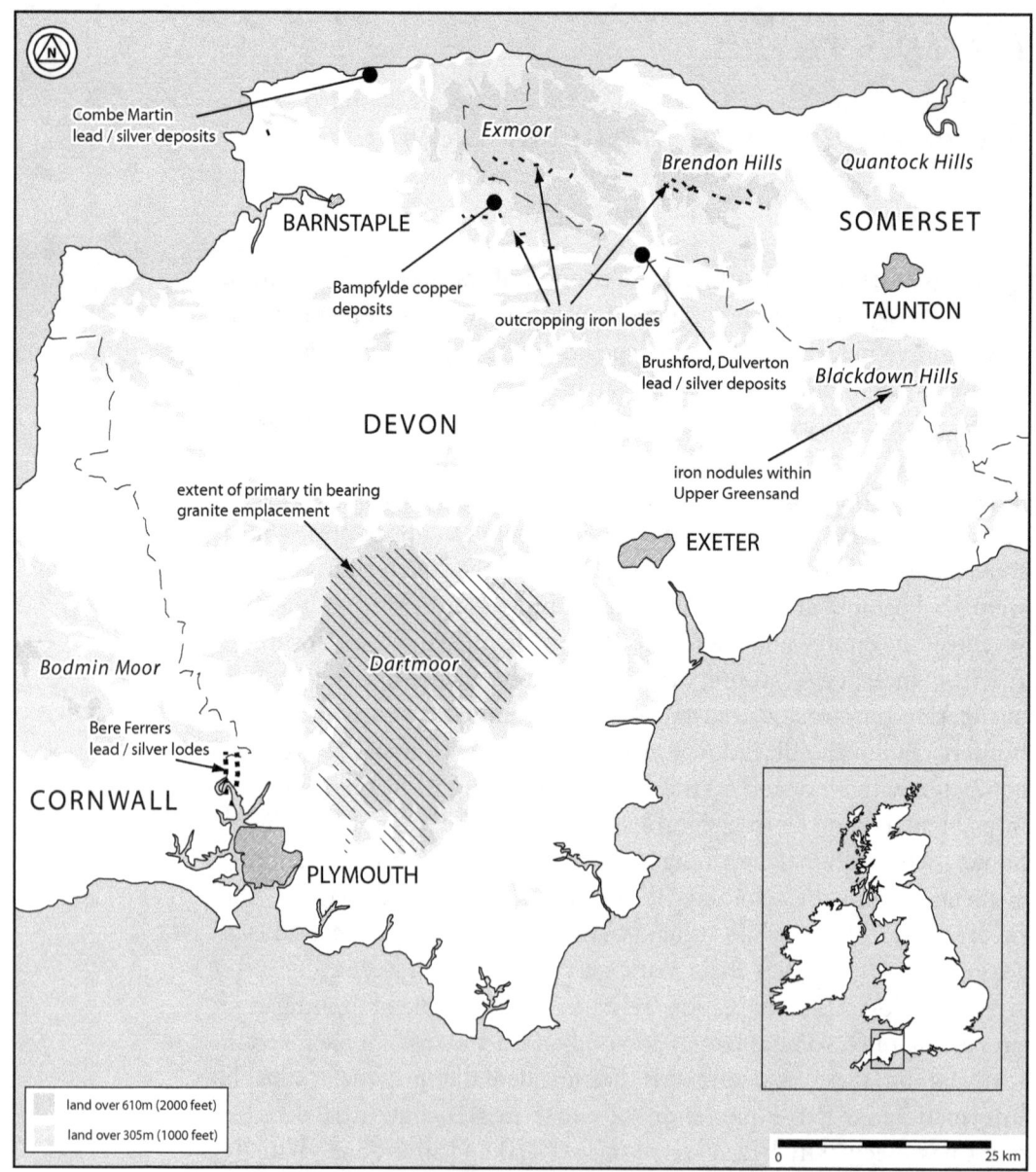

Figure 2.1: The major mineral resources of Devon and adjacent areas (information taken from Dines, *The Metalliferous Mining Region of the South West*, vol. 2, map XIV; Hawkes, 'Dartmoor Granite', fig. 5.6).

in South-west Britain was published, our knowledge of the archaeology of earlier mining in the region was limited.[2] Tin-working had perhaps received the greatest attention with work by Greeves and Gerrard, but the detailed recording and mapping of the industry by the Royal Commission on the Historical Monuments of England, now part of English Heritage, was in its infancy.[3]

Since the 1990s it is the study of iron, and particularly iron smelting, in Devon and adjacent parts of Somerset which has been the focus of archaeological investigation, although work has also been carried out on lead/silver working at Combe Martin in the north of the county. Griffith and Weddell, and a series of the University of Exeter's 'Community Landscape Project' reports have examined aspects of the iron industry on the Blackdown Hills[4] while the University's 'Exmoor Iron Project' has added significantly to our understanding of the industry in that area.[5] Recent archaeological work at Combe Martin has highlighted the strength of lead/silver production there in the immediate post-medieval period and has identified potential sites of early (possibly medieval) activity (e.g. Figure 2.2).[6] The majority of the research on lead/silver has, however, been based on the documentary evidence for working at Combe Martin and Bere Ferrers,[7] and documentary sources have also fuelled the work carried out on the history of the tin industry from the twelfth century onwards, particularly the pioneering work by Finberg and Hatcher.[8]

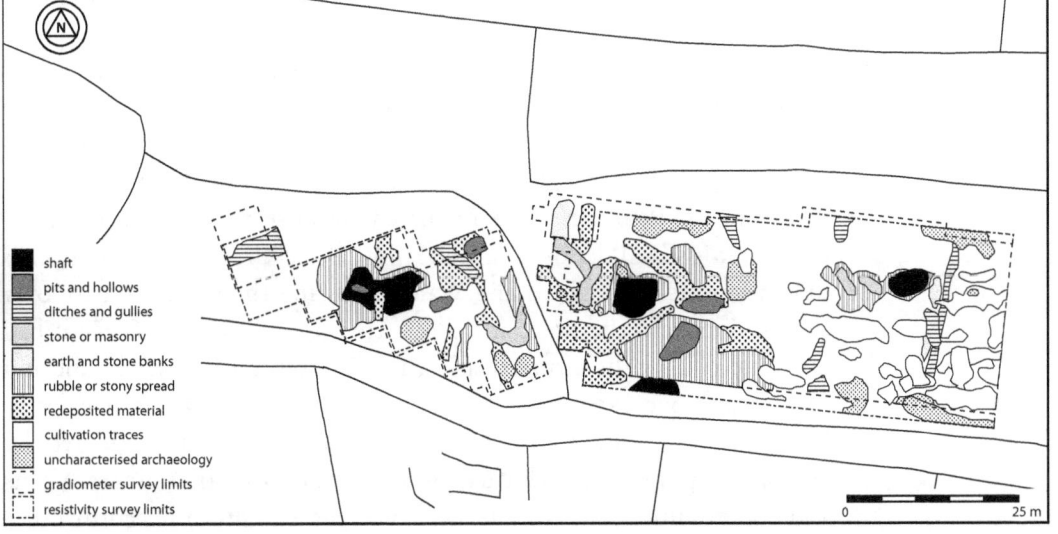

Figure 2.2: Combined interpretation of gradiometer and earth resistance surveys of workings on Harris's Lode north of Corner Lane, Combe Martin (after Substrata Ltd *Gradiometer and Resistance surveys at Mine Close*, fig. 1).

Mineral ownership in the medieval period

Before we move on to examine the metal resources of Devon in more detail, we should first consider their place in English law. There is a general perception today that, in contrast to the rest of Europe, the ownership and the right to work minerals in England, and the rest of the United Kingdom, lay with the owner of the land. Even today the position is not so simple, and it was certainly far more complex in the medieval period.[9] In fact, the late medieval was a period of change as the state (the English Crown) and those working the minerals sought to establish their rights to valuable resources. Understanding ownership and the regulation of mining are important factors in interpreting both the development of mining and its impact on the landscape.

For our purposes, the minerals in Devon can be divided into three groups defined by ownership and the rights accorded to those working them. The first group includes those metals over which the English Crown exercised a right of prerogative from the mid-thirteenth century onwards – silver (along with the lead with which it was associated), gold, and copper. Secondly there was tin which, although it was the property of the landowner, was subject to customary rights allowing it to be worked by any miner or 'tinner' who paid a portion to the owner and a duty or tax (coinage) to the Crown which also had a right of pre-emption on the metal produced (that is the right to buy the produce at a set price). This leaves a third group, the other minerals including iron, where ownership was vested with that of the land; where there were no customary working rights, or they are now lost; and where the Crown had no call on the produce.

The prerogative over silver, gold, and copper was first exercised in the 1260s in response to potential discoveries of those metals in Devon.[10] Prior to that both silver and gold could, like lead, be worked as a right in common.[11] Silver-bearing deposits on the Crown demesne in the north Pennines were regulated by custom, as were the non-argentiferous lead deposits on the lands of the Crown and lesser lords in other parts of England and Wales, allowing the miner control over the produce and the time, and place, that it was worked. Once it had introduced its right of prerogative, the Crown, on the whole, did not seek to remove existing customary rights. In the sixteenth century it sought to extend its prerogative into mining fields regulated by custom, but was then rebuffed by the strength of the miners' rights and the common law. It was not until the seventeenth century and the growth in commercial interests among the new land-owning gentry that customary regulation of mining was eroded.[12] However, in Devon in the late thirteenth century, there is no evidence for existing working of the silver-bearing lead deposits and the Crown was able to take control of those deposits and introduce direct management of the mines without reference to custom. The practicalities of controlling the extraction of gold are discussed below, but it, and the copper deposits in the north of Devon, were also subject to Crown control.

Customary rights to work tin in the south-west of England, like those for lead in other parts of the country, were in a form which probably pre-dated the Conquest in the eleventh century. It is evident that the English Crown had established an interest in tin production by the middle of the twelfth century when the collection of a 'farm' (a lease on the income due) on the duty on tin was recorded in the Pipe Rolls.[13] The 'Stannaries', as distinct tin mining areas in Devon and Cornwall under Crown control, are first documented during the second half of that century. An inquisition of 1198 regularised the arrangements for payment of tax (coinage) on tin and confirmed the existence of mining customs within the Stannaries.[14] Those customary laws were codified in a charter of 1201 and allowed the miners (tinners) liberty to search for, and work, tin without interference, to divert water as required and take fuel for smelting. Tinners were to answer only to the Warden and his officers who were granted exclusive jurisdiction over them. The customary laws were subsequently subject to modification, notably by the Charter of 1305 which granted tinners immunity from general taxation, and regulated tin mining until they were incorporated into statute law in the nineteenth century. Where deposits were found on its demesne, the Crown was the mineral owner, but in other areas the tin remained the property of the landowner and the tinner was to pay a portion of the produce to them. The Crown confined its involvement in the industry to overseeing regulation, exacting tax (coinage) and, on occasions, exercising a right of pre-emption on the tin produced.[15]

The ownership of iron deposits in England and Wales has always been linked to land holding. In only one area, on the Crown demesne within the Forest of Dean in Gloucestershire, is the mining of iron subject to regulation according to custom. It is possible that iron mining was subject to local custom in the south-west of England during the medieval period but, if it was, the evidence is now lost. The place and manner of working was under the control of the land owner and subject to agreements drawn up between the owner and the miners. Unfortunately, no such agreements survive for Devon in that period and we are reliant on limited secondary documentary, and archaeological evidence for information on the way that the mines were worked.

The metal resources of Devon in their national context

Metal resources played a significant role in the development of society and commerce in late medieval England. The precious metals gold and silver, in particular the latter, were the basis of an expanding cash economy. Although most of the silver came from continental European sources, newly mined English silver was still important in establishing a high quality coinage in the twelfth century and supplementing imported silver thereafter. Production of tin and, to a lesser extent, lead contributed to the export earnings that drew in continental silver. Lead, iron and, to some extent copper, had essential domestic and martial uses which governed their exploitation. Metalliferous mineral occurrences and the mode of their exploitation, together with the historical and archaeological evidence are discussed below. The sequence in which they appear is ordered by ownership and the rights governing their working, the division of which was outlined above – silver and lead, gold and copper, followed by tin and iron.

Lead and silver

Silver-bearing deposits in the south-west of Devon, in the Tamar valley at Bere Ferrers, are the principal subject of this book, but they were not the only sources of that metal in Devon or its borders. Virtually all the silver found is associated with lead ores. The occurrence of silver in copper ores is dealt with below as are the more complex silver-rich ores found in East Cornwall. There was prospecting for silver at various locations in the south-west of England, but it was only in the north of the county, at Combe Martin and to a lesser extent near Dulverton on the Somerset border, that there were other significant workings for silver in the late medieval period.

Despite prospecting carried out by the Crown over the previous thirty years there is no documentary evidence for any lead/silver production in Devon before the opening up of the Bere Ferrers and Combe Martin mines in 1292. However, the simultaneous opening of two groups of mines at the southern and northern extremities of the county does suggest some prior knowledge of the silver-bearing deposits. With no existing mining interests to displace, the Crown could employ direct management of the mines without reference to custom. The workforce, many of whom were immigrants to the area, some being pressed into service, were employed directly by the Crown and were subject to the control of Crown officers. Within six years there were three to four hundred miners at work, although by that time the mines at Combe Martin had been abandoned by the Crown and work was concentrated at Bere Ferrers. The composition of the workforce tells us something of the task they faced in mining and smelting the silver-bearing ores.

Despite the excellent detail in the documentary sources regarding the mines, their operation, production, and staffing, there is little reference to how the mines and miners engaged and operated with and within the agricultural landscape, and particularly where the miners lived. After most of the smelting and the refinery were moved to Calstock some accommodation was provided: 'houses for the workers spending the night in the *curia*'.[16] This does suggest that their homes were some distance away, perhaps closer to Maristow, and it is in that area that the search for at least some of the associated settlement should focus. The keeper of the mines, and at times the controller, was resident at the site of the refinery, firstly at 'Biccombe' and when smelting operations moved in 1301, at Calstock. Unlike mining itself, which, until the introduction of adequate drainage technology would have ceased due to winter water levels, smelting and refining may have been an almost year-round activity.

It is a possibility that during winter months miners returned to their home counties, perhaps to their families, and so the year-round demand for food fluctuated with the mining seasons. As the introduction of adit drainage made it possible to mine in the winter, it is possible that miners may have taken tenancies and partially existed under a system of 'dual-occupation', although remaining principally miners.[17] However, Hatcher's and Blanchard's model of those involved in 'dual occupation' is someone who is neither year-round miner nor year-round farmer (for a more detailed discussion of this model see 'Tin' below). As such, involved in a year-round mining calendar, it is difficult to picture the miners of Bere Ferrers as being involved in farming. Supporting this notion, it is known that miners were impressed from other British mining regions, 150 in 1297 for example. These men would have had no existing tenurial rights over the surrounding land, and unless large areas of unfarmed land were brought into use, such as areas of common being allocated to miners, or existing holdings were subdivided, the opportunity for the incomer to farm his own land was limited, at least until their permanence in the locality was established. Continuity within the mining population certainly does receive some support in the wage rolls, with the same family name occurring over a number of years, suggesting established family units and this is discussed further in Chapter 4. Similarly, Derbyshire surnames are to be found in the mine's account roll of 1480/81, and although it was still possible for lessees of the Crown to impress miners, as the Crown had done up to 1350, this may also suggest a family lineage from earlier generations of mining families. By the mid-fifteenth century demographic decline was such that the chances of persuading miners to move to Devon would have been slim, and by this time the mines were probably staffed at all levels by local people.[18]

The close association of silver with lead deposits meant that the silver mines were an important producer of lead as a by-product of the refining process. By the late twelfth century, the non-argentiferous lead deposits of Mendip, the Peak District of Derbyshire, further north in the uplands of Yorkshire, Durham and the North Pennines, along with west Shropshire and north-east Wales, had developed as major sources of metal for fabrication purposes, supplying the demand created by an expanding martial and ecclesiastical building programme. The silver mines in Devon were well placed to contribute to that market. Lead from the south-west of England has been identified as far a field as St Davids, in the south-west of Wales, where it was used in the construction of the bishop's palace.[19] The method of provenance, lead isotope analysis, can only link it to the mineralisation associated with the granite emplacement in south-west England, and so by excluding the earlier mineralisation at Combe Martin, and with a construction date in the late medieval period, it does suggest that the lead was a by-product of silver refining at the Bere Ferrers mines. Between 1304 and 1317 those mines sold at least 582 cartloads (around 428 tonnes) of lead. Some was used locally, for example, in the construction of Exeter Cathedral and when that was not available, as in 1302, it was sourced at the fair in Boston, Lincolnshire, where lead from Derbyshire was available to a wider market.[20]

Prior to the late twelfth century the demand for lead had to a large extent been satisfied by the output from silver refining, with the mines of the north Pennines providing the bulk of production.[21] Some 235 cartloads (around 173 tonnes) of lead were sequestered by the Crown from the mine of Carlisle in 1181, largely for the use of the abbey at Clairvaux.[22] This was the last record of significant lead production from the silver mines in the north Pennines, and by the late 1180s those mines were in decline. Non-argentiferous producers in Yorkshire, and elsewhere, were providing the bulk of the lead required by the Crown for its construction programme and that of the ecclesiastical institutions of which it was a benefactor. Between 1179 and 1183 at least thirteen shipments of lead were made from non-argentiferous sources in Yorkshire to the king's works along with ecclesiastical houses in southern England and France.[23] The Shropshire mines were also active for the same purpose at this period, and earlier if the unspecified mine let to Drogo *minetar'* in 1162/63 was worked for lead, with metal being purchased by the Crown and shipped down the Severn from at least 1179.[24] From a similar date there had been shipments from mines in the Derbyshire Peak District and these continued into the thirteenth century when a market for lead was established at Boston.[25] The castle building programme of Edward I, particularly that begun in Wales (1277–95), gave a boost to lead production. Building the Welsh castles drew in lead from north-east Wales, from Shropshire, Derbyshire, Mendip, and from the Isle of Man.[26]

The majority, if not all, of these mines were governed by customary law which gave the miners control over production in return for a payment of a portion of the produce to the lord of the soil.[27] The use of custom to control manorial resources was commonplace in the medieval period. Where mining, particularly mining for lead and tin, differed from most other resources was that it was carried out for the monetary profit of both tenant and lord. Lead mining provided a cash income for the miner, but he still relied on his agricultural holding to provide a living.[28] By the end of the thirteenth century, however, the linkage between mining and agriculture had been weakened by population growth and a shortage of land. This population boom provided a pool of labour in the lead mining areas on which the Crown could draw for the silver mines in Devon.

Production of lead was largely for domestic consumption, with many ecclesiastical institutions becoming involved in mining, but there was also an export element. Quantifying that production is extremely difficult. As Miller and Hatcher point out, the evidence is unsatisfactory and incomplete.[29] Blanchard suggests the Derbyshire production stood at not less than 391 fothers (about 373 tonnes) in 1300 but this masks a variable record over the previous hundred years.[30] As output from one part of the mining field fell others rose to compensate, in part at least. This was to be a feature of Derbyshire and other lead fields through the late medieval period. The mines, known as 'grooves or groves', were shallow linear openworks or shallow shafts exploiting fissure veins, 'rakes', and associated replacement deposits, or 'pipes', in the limestone. As these veins were worked out to the water-table and all suitable ore exhausted, mining and smelting activity migrated to new deposits. In this way production kept pace with demand without putting any great technological or financial burden on the miners. Production from Derbyshire peaked around 1300 before declining again during the first half of the fourteenth century, a trend which, over the previous 150 years, had mirrored the demand from major building works.[31]

As with other mining activity, lead production suffered from the effects of the Black Death. Some Derbyshire mines were abandoned. In north-east Wales production from the Minera field appears to have ceased altogether and the fines paid on the Englefield mines were greatly reduced 'because a greater part of the miners were dead, and those that survived were unwilling to work'.[32] This suggests a parallel with tin production in the south-west of England, where miners appear to have abandoned mining to concentrate on working vacant agricultural holdings.[33] Derbyshire showed signs of recovery by the 1360s, as the quantity of tithe ore for Hope, Bakewell, and Tideswell started to rise and continued to do so to 1403.[34] There then followed a period of fluctuating output, but the overall trend to mid-century was one of decline until it stagnated at very low levels.[35]

When recovery began, after c.1475, it did so largely to feed a new market as the continental European silver producers adopted new technology to extract silver from copper ores using a form of liquation, the saiger process. This process consumed lead as it was added to partially reduced copper ores to combine with the silver and was then refined by cupellation. The domestic market for new lead remained depressed well into the sixteenth century due largely to the availability of material recycled from redundant ecclesiastical buildings. Lead production, largely for export, had risen to over 650 tonnes by 1508. Some of this increase in the 1450s came as a by-product of the Devon silver mines, but, as producers in Derbyshire, Yorkshire and Durham and, by the 1490s, Mendip reacted to the demands of the new market, output rose across the country.[36] Of the lead fields active prior to 1400, only north-east Wales failed to respond. Production there had been adversely affected by the suppression of the Glyndwr rising, from 1400 to c.1413, such that as late as 1472/73 the bailiff at Minera accounted

> for the losses and reduced rents from divers lands and tenements ... and other profits accruing to the lord from the mines which from ancient times until now have been paid by them, which several tenements and houses lying there were destroyed during the attacks and incursions made by the lord King's rebels from parts of Wales and have not been repaired because of the absence of many tenants and because of the poverty of the same tenants and up to this time have lain in the lords hands.[37]

Mines to the north in Flintshire were farmed out (let at a fixed annual rent) in 1423 for a mere £3 18s. 9d. and in 1509 those at Holywell and Vaynol failed to yield a profit as no one was willing to work them.[38]

The mines of north-east Wales did not see renewed production of any substance until the seventeenth century. By that time the customs by which they had been governed had fallen into disuse, allowing the lord of the soil to gain full control over-production.[39] Output from other lead fields, in Derbyshire, Mendip, and parts of Yorkshire, continued to grow after 1500 and survived the depression in prices caused by the release of large amounts of monastic roof lead onto the market in the 1530s. Their customary law also survived, regulating production in some cases until embodied in statute law in the nineteenth century.[40] As new centres of production developed outside the lead fields governed by custom during the early modern period, they did so on a scale not available to the medieval lead industry. The land owners and entrepreneurs working deposits in mid-Wales, Shropshire, and the resurgent fields of the north Pennines and north-east Wales from the seventeenth century onwards controlled both the produce and the means of production, allowing them to make full use of external finance for technological developments.[41]

The field evidence for lead mining in England and Wales in the medieval period is extensive, albeit intermixed with that for the post-medieval period.[42] In some mining fields, particularly those within the Peak District and Yorkshire Dales National Parks, the landscape impact of mining has been assessed[43] but for many fields, such as Mendip which was a key producer and source for innovation, even such basic work has yet to be carried out. The landscape of lead mining, the working methods, and their relationship to the pattern of settlement, provide for comparison and contrast with that of silver mining in Devon, with the latter drawing to a large extent on lead mining for its workforce and its hard-rock mining techniques.

Copper

The archaeological and documentary evidence for copper working in the British Isles prior to the late sixteenth century is scant, and much of the evidence is specific to Devon, and to the late medieval period.[44] From the mid-thirteenth century copper, along with silver and gold, was included in the Crown prerogative despite its limited economic value. Copper was used in Sterling silver coinage, but only in small quantities – no more than eight per cent by weight. Copper's role as a roofing material in central Europe was largely filled by the use of lead in England and Wales, although it was a significant component in the production of both brass and bronze. The latter had only limited use prior to the introduction of ordnance in the fifteenth century, after which bronze was used for the manufacture of cannon until it was gradually replaced by cast iron from the late sixteenth century onwards. Even then the English government found it preferential to source copper from the dominant producers in Sweden rather than exploit local resources, despite the risk of disruption to supply during periods of European conflict. This danger was well illustrated by events of 1540–41 when an embargo was placed on up to 4,000 cwt of copper, paid for at Antwerp, intended for Henry VIII's campaign in the Low Country.[45]

Copper ores are found in both north and south-west Devon. In the north of the county the copper at North Molton (Figure 2.3), and elsewhere along the southern border of Exmoor, occurs alongside iron in chalcopyrite-siderite ores. These were probably of hydrothermal origin, showing some evidence of fault fissure filling, and follow the cleavage of the rock with a general east-west alignment. The north Devon mineralisation appears to be earlier than that in the south-west of the county, which is associated with the granite emplacement. The copper ores found as fault fissure deposits around Dartmoor and into the Tamar valley are not as rich as those on the Exmoor fringes, although they were far more extensive.[46]

The lack of evidence for medieval and earlier extraction and processing prompted Barton to suggest that British copper mining did not really take hold until the late sixteenth century and then it was something of a false start with activity in Cumbria foundering on the economic reality of a weak market for the metal.⁴⁷ It was not until the last decade of the seventeenth century and into the eighteenth century that English copper developed as a strong industry with Devon at the forefront and attracting the attention of foreign competitors.⁴⁸ In south-west Devon the depth of the fault fissure deposits rendered them difficult to identify, and also to exploit, and as such were probably not the earliest in the county to be mined. Conversely, it is not surprising that there is a growing body of evidence suggesting that North Molton ores were worked during the medieval period. They are rich in copper with up to 50 per cent metal in the oxidised ores at a relatively shallow depth. The lack of glacial action in the last Ice Age meant that that zone had not been eroded and was followed by an unusually deep enriched zone below the water table, down to c.200 metres in depth, which was worked from the 1690s through to the nineteenth century.⁴⁹

There have been suggestions that copper was mined at North Molton by the 'Romans' and later during the reign of King John,⁵⁰ but the earliest documentary evidence is for the mid-fourteenth century in a licence granted to Nicholas de Welliford to work the king's copper mine at North Molton, although this was quickly rescinded when it proved unprofitable.⁵¹ However, as Dixon has suggested, the grant, and the quick realisation that no profit might be made, could imply the reopening of an existing mine.⁵² The mine at North Molton was again referred to as the king's mine in a survey of c.1524 by which time, as

Figure 2.3: The Bampfylde copper mine in North Molton and associated features in relation to the late nineteenth-century landscape (based on the Ordnance Survey First Edition Six Inch map). Note the 'Early Mines?' which are possibly the most tangible location of medieval mining and what Rottenbury (*Geology, Mineralogy and Mining History*) referred to as 'Elizabethan mines'.

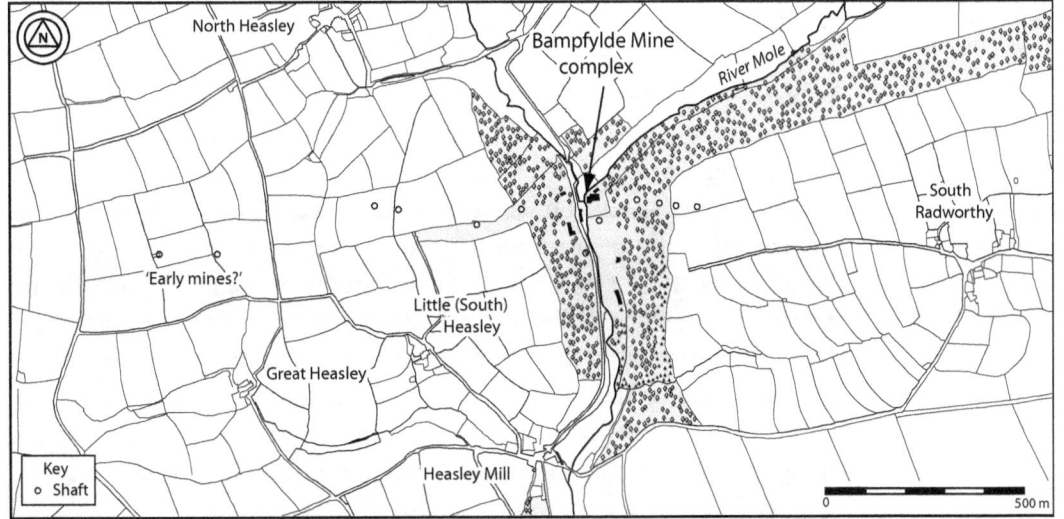

is outlined below, there is archaeological evidence suggestive of on-site smelting.[53] As yet, none of the surface features at Bampfylde (NGR SS737327) (see Figure 2.3), the most likely site of the king's mine, can be dated to the medieval period. There is an adit on the western bank of the river Mole and shallow workings at Great Heasley (the latter associated with slag finds), to the west of the Bampfylde Mine, both of which are said to be hand cut, and so may predate the 1690 workings. Unfortunately they are no longer accessible and the only evidence that survives are notes made by geologists in 1919 and later.[54] Those workings may predate renewed mining in the late seventeenth century, but, although explosives were in use at Bampfylde in the 1690s, they were expensive and hand cutting techniques continued in use where possible.[55]

Finds of copper smelting slag at the Bampfylde Mine have been reported since the nineteenth century, although there is no documentary evidence to support the smelting of copper on the site.[56] There is, however, ample evidence for the shipping of the ore for smelting from north Devon to Bristol, the Wye valley, and south Wales, from the late seventeenth century onwards.[57] However, copper-rich slag with entrapped charcoal, recently found at Bampfylde, has been radiocarbon dated to the late fifteenth century,[58] and almost certainly indicates the local smelting of copper ore, although the site of the furnace has yet to be identified (North Molton is too far inland for the slag to have been brought in as ballast on a ship returning from one of the Bristol Channel ports).

The copper ores of Dartmoor and the Tamar valley, as at North Molton, were not worked on a large scale until the end of the seventeenth century. By the mid-nineteenth century the output of copper from Devon Great Consols, about five kilometres west of Tavistock on the east side of the river Tamar, had reached world-dominating proportions. There is, however, good documentary evidence that some copper-bearing ores were being worked in the Tamar valley by at least the first decade of the fourteenth century. Smelting of those copper ores was undoubtedly being carried out for their silver content and evidence for the recovery of the copper comes from the residues of that process (see below). It is also probable that the attraction of the North Molton copper ores was the silver or gold content, the former of which ranged from five to nine ounces per ton in ores found in the enriched zone and was probably significantly greater in the oxidised zone closer to surface.[59]

Despite the evidence for the smelting of copper ores in the Tamar valley in the fourteenth century, the site where they may have been mined remains elusive. There are nineteenth-century references to early working in some copper mines, such as Wheal Gatepost (later known as Devon Burra Burra), where it was said that 'much work had already been done by the "ancients"'.[60] Most, if not all, of these references might be explained by the extensive prospecting carried out in the wake of the expansion of copper mining in the eighteenth century. Hard archaeological evidence is lacking, although it should be pointed out that the detailed underground survey required to identify such early workings has yet to be carried out in the south-west of England.

Gold

The occurrence of gold in Devon and Cornwall has from time to time attracted the attention of mineral owners and the public. There are, given a sufficiently high gold price, sizeable low-grade deposits which might be worked at a profit, but for which there is no historical or archaeological evidence for working prior to the advent of modern geological prospection techniques.[61] Gold is, however, found as a component in the placer deposits worked for their tin content and in the weathered, oxidised, zone close to surface on some copper deposits, notably those along the southern border of Exmoor at North Molton and Molland.[62] The former were certainly known in the medieval period and it is probable that the latter had attracted the attention of the Crown by the thirteenth century.

Gold has been worked in England and Wales since before the Roman occupation. Normally it would have been recovered from placer deposits along with tin, or as chance finds of gold eroded from primary deposits. The exception was the Ogofau mine at Dolaucothi, near Pumsaint in Carmarthenshire, where existing gold mines on a reef deposit appear to have been re-established by about AD 74. Recent archaeological investigation indicates that new workings and an extensive water supply system were superimposed on existing opencast pits, and shallow underground workings were extended to a depth of up to 30 metres below surface.[63] After the Roman withdrawal there is no hard evidence for the mining of primary gold deposits in England and Wales until the nineteenth century.

The right to work gold-bearing deposits was regularly recited in grants made by the English Crown from the early fourteenth century onwards and there is specific reference to its occurrence in tin-workings. In 1325 22 dwt of gold was recovered from mines in Devon,[64] and in 1377 Henry de Burton produced gold 'found in a river in Devenshire [sic]', whereupon he was tasked with searching out further deposits.[65] Such gold was probably the product of tin streaming activity, as described in Cornwall in later periods,[66] but there were no subsequent reference to payments in gold into the Exchequer. The tinners were, no doubt, adept at concealing the gold and paying only coinage on their tin production. Returns for 'St Tether', probably the parish of St Clether on the north-east border of Bodmin Moor in east Cornwall, in the period 1445 to 1451 record the production of silver worth three marks but state that no gold was found.[67]

Any suggestion that gold might have been recovered from primary deposits is confined to the possibility that the gold/silver/copper deposits which attracted Crown interest in Devon in the 1260s, and which led to the Crown initiating its right of prerogative, lay on the south borders of Exmoor. Our only information as to their location is that they were found at 'La Hole' and that those investigating the deposits were granted the use of wood for charcoal production from the king's forest of Chittlehamholt, centred to the south of South Molton in north Devon. The forest in the thirteenth century stretched north-eastwards towards Molland and, although 'Hole' is a common place-name element in Devon, there are a cluster of tenements ending in 'hole' close to the copper deposits in that parish. The Molland copper deposits do carry some silver and were considered to be gold bearing in the mid-nineteenth century.[68] Although they were never exploited in the same manner as the North Molton deposits in the nineteenth century there is the possibility that gold occurrences there were the subject of Crown interest in the thirteenth century.[69]

Overall, gold, like copper, was of minimal economic importance in Devon during the medieval period. Its attraction, including use in coinage from the 1340s, stimulated the continued search for the metal, but its impact on the landscape is limited, wrapped as it is in the evidence for tin extraction.

Tin

Of all of the mineral resources exploited during the medieval period in Devon, tin was the most extensive industry and of the greatest economic significance. Tin-working has also seen the most intensive archaeological and historical study of any of Devon's metal industries.[70] The work of the Dartmoor Tin-working Research Group, and that on-going with English Heritage, instigated the detailed recording and mapping of the industry.[71] To date little scientific dating has taken place, and the excellent documentation of the Stannaries has been relied upon to date mapped tinworks. Recently, however, analysis of sediments has been used to date and assess the scale of tin-working within the catchment of rivers to the south-east of Dartmoor.[72]

The English tin industry was already worth over £7,000 per annum by the early fourteenth century, of which Devon contributed over ten per cent, and by the end of the fifteenth had risen to some 600 tonnes per annum worth at least £15,000 on the London market. This was achieved using undemanding techniques to exploit relatively shallow deposits in an industry which was still open to the small operator. It was only at the end of our period that tin was first extracted directly from its primary source, the cassiterite- (tin oxide) bearing veins or lodes found either in the granite or the adjoining metamorphised rocks from Dartmoor in the east across Cornwall to Penwith in the far west. Up until the fifteenth century the source was almost entirely secondary 'placer' deposits composed of material eroded from the tin lodes and deposited on the slopes of the granite uplands or in the valleys which surrounded them. Most Devon tin-working of the twelfth to sixteenth centuries focused on rich alluvial deposits in the valleys around Dartmoor. These contained between 0.022 per cent and 0.15 per cent cassiterite which, after the waste 'gangue' material (the clays, gravels etc.) had been removed, was commonly referred to as 'black tin'.[73]

The demand for tin during the late medieval period attracted external finance, largely provided by local men of substance, merchants, gentry, and clergy, keen to secure a share of the produce. Working capital was required to finance tin-working, a labour intensive activity employing a significant number of full time workers. With producers unable to sell tin before the infrequent coinages, tinners were obliged to take loans, guaranteed against future production, to maintain their activity and so from an early date the tinner was 'enmeshed in a web of indebtedness'.[74] The money advanced to the tinners by local men came in part from London and alien merchants. There was what Hatcher describes as a 'three tier' system financing production and securing the tin presented for coinage: tin dealers, merchants, and miners. Investments were also made in tin-working as merchants and others became partners in ventures, providing the capital with a view to sharing the profit.

Hatcher provides an outline of the diversity within the tin industry, where the working structure, that is the ownership and running of the claims, varied from lone prospectors, partnerships of tinners (some partners may have substituted hired labour or financial recompense for their own personal toil), to single or groups of merchant tinners who used hired labour to work their claims. Labouring tinners may have worked full or part-time, either on a permanent or casual basis. Wages may have been a fixed cash sum or piecework, a proportion of the product, or a combination of cash and product.[75]

In contrast to other minerals like lead, found in veins or lodes, extraction sites on the placer tin deposits were customarily defined by area. Extraction of linear deposits, such as those for lead, were defined by length (by 'meers' of varying length) with only a limited amount of ground, a matter of a few metres, on either side of the deposit being allowed for surface operations such as ore processing and the disposal of spoil. The tinners, however, were expected to mark out the limits, or 'bounds', on a working area of approximately one acre (0.405 hectare) and, from at least 1494, they were to register them with the appropriate Stannary court, of which there were four in Devon: Tavistock, Chagford, Ashburton, and, from 1328, Plympton.[76] When, by the fifteenth century, the primary lode deposits were being worked for tin, the form of 'bounding' by area was retained giving the tinners the advantage, over contemporaries in the customary lead mining fields, of being able to work all the deposits within their bounds and to utilise a much wider area for surface operations.

The exploitation of placer deposits is collectively known as streamworking, although the method varied depending on whether they were alluvial (valley bottom deposits) or elluvial (found on hillsides in close proximity to the primary source). Once identified by digging prospecting pits or hatches, the working of the alluvial deposits involved removing the covering layers of peat and fine-grained sediments to expose the cassiterite-bearing placer gravels.[77] Cuttings were generally, but not exclusively, made from the lowest point, working upstream or upslope. The tin ore was separated from the gangue material by a controlled process of washing, using either water from adjacent streams or water carried to the workings by a leat, the right to which was an established custom and incorporated in the earliest charter.[78] A proportion of the gangue material inevitably entered the streams, and had an effect on downstream sedimentation rates.[79] This had a contemporary impact, with harbours such as Plymouth and Teignmouth suffering silting beyond the norm, significant enough to demand an Act of Parliament of 1531 aimed at reducing the amount of tinning-derived sediment entering river systems.[80]

There was a need to maintain a sufficient velocity of water to wash away gangue material and maximise the amount of ore left behind. In alluvial streamworks this was done by regulating the angle of the work area (the tye) or the amount of water running across it. A regular width

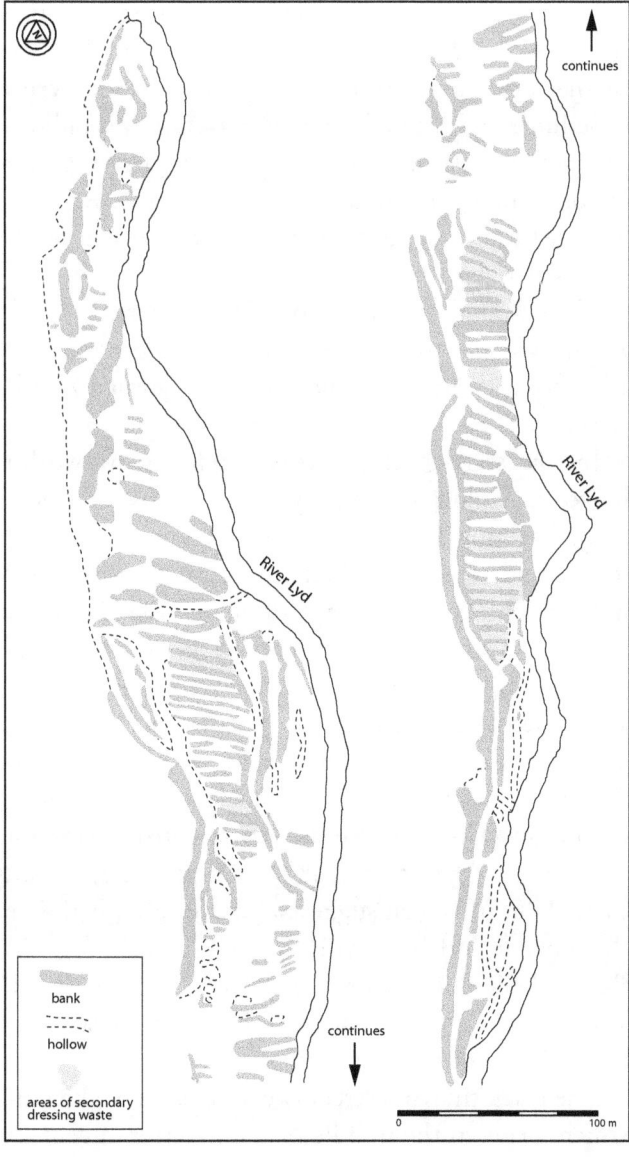

Figure 2.4: The tin streamworks at Lydford, south-west of the Dartmoor granite emplacement (after Gerrard 'The early south-western tin industry', fig. 22).

of tye was created using dumps of spoil, resulting in the characteristic morphology of these sites, as witnessed adjacent to the river Lyd in Lydford Woods to the north of Tavistock (see Figure 2.4). On elluvial deposits the workings were orientated at angles across the hillside, sometimes served by a number of leats, thereby managing the velocity of the water across the tye, with examples surviving at Beckamoor Coombe and the northern side of Newleycombe Valley.[81] Where water was not readily available at some of the higher eluvial sites, the deposits might be worked by pitting and processed elsewhere.[82]

Increased sedimentation caused by tinning has left a geoarchaeological signature that has been used to identify and date areas of tin-working further upstream.[83] In the Erme valley, in southern Dartmoor, a layer of sediment with a high tin concentration was radiocarbon dated to cal. AD 1267–1397, a date obtained from underlying organic-rich silt, and is concurrent with documented tinning activity on Dartmoor.[84] An earlier, thinner deposit suggests less intensive or short-lived tinning, and is dated to between cal. AD 245–366 and cal. AD 460–730 (i.e. the late Roman or early post-Roman period).[85] In the neighbouring Avon valley tin-rich sediments have been dated to cal. AD 1448–1621, coincidental with the sixteenth-century zenith of Dartmoor tin streaming.[86]

By the start of the sixteenth century tin-working in Devon had moved to exploit the primary lode deposits. The most common method of prospecting for these was by digging a line of pits at 90 degrees to the expected lode. Archaeologically, these are seen as a circular hollow with a spoil heap, often on the downslope side, although there are also examples of hand-dug, and sometimes water scoured, prospecting trenches.[87] The latter are not known in Devon, but there are numerous examples of prospecting pits, for example those on Black Tor.[88] Once identified, the primary deposits were worked by relatively shallow 'lodeback' pits or, where feasible, as deep linear openworks or 'beamworks'.[89] There are no scientific dates for the exploitation of primary deposits or for stream-working, although the fieldwork carried out in Devon suggests that the latter is the earlier. Greeves sees the move to lodeback working as a logical progression and, based on the available documentary evidence, suggests that deeper working by shafts was in progress at Furzehill, a tinworks near Horrabridge, by the early sixteenth century.[90]

The black tin from streamworks could be taken directly for smelting which, prior to the middle of the fourteenth century, was carried out in small shaft or bowl furnaces.[91] It was a two-stage process, a first smelting on or near the workings and a second smelting (refining) at Exeter or another town designated by the Warden of the Stannaries.[92] Both smeltings were initially points at which tax was levied, but by 1305 coinage was divorced from smelting. Subsequent improvements in smelting made the second smelting unnecessary. The larger furnaces with water-powered bellows, blowing mills (in Devon – blowing houses in Cornwall), were then capable of producing metallic tin in one process.[93] A number of blowing mill sites are known on Dartmoor, but none has been dated to the medieval period. No sites for the earlier furnaces are known in Devon although what is believed to be an early thirteenth-century site has been investigated at Crift Farm, near Lanlivery, in Cornwall. Analysis of the slag, and comparison with that from known blowing house/mill sites, showed no real differences that might be used to identify early tin smelting sites.[94] The demand on natural resources by tin smelting is in evidence across much of Dartmoor where peat deposits have been worked to supply the 'moor coal' (peat charcoal) required as fuel.

Tin ore extracted from primary deposits required processing before it could be smelted. This involved physically breaking down the ore to release the lighter waste or gangue which was then removed by gravity separation in a controlled stream of water (dressing), processes which, in the stream-working of alluvial deposits, had largely been carried out by nature. The 'tin mill', a water-powered set of stamps designed to effect the process of breaking down the ore, appears to be a product of the move to working the primary deposits although the process would have been required, on occasion, to treat the courser alluvial deposits. The known tin mills, and the associated dressing floors, are therefore largely assigned to the early post-medieval period.[95]

The pressure of population growth on the settlement of Dartmoor and its borders, particularly post-deforestation in 1240, might have been expected to bring tinners into conflict with agricultural tenants and that it did not is seen as an indication of the involvement of the latter in tin-working. Mineral extraction was perhaps seen as the best use of marginal land, even that which had been enclosed as pasture.[96] There was, nevertheless, a certain separation between the larger settlements around the edge of Dartmoor and the tinworks on the moor and in its valleys. By the early sixteenth century, when extensive records are available, the involvement of a wide cross-section of society is attested.[97]

Gerrard provides a discussion of the issues surrounding the places where tinners lived, critically observing that as yet, no tinners' homes have been identified.[98] A key consideration is whether tinners were engaged on a full-time basis and what their position within the wider agricultural community was. Hatcher and Blanchard have actively debated this, with Blanchard believing that land-owning tinners (or lead or iron miners) were self-sufficient in terms of food, living as 'husbandman-cum-miner[s]'.[99] In reconstructing the calendar of such a person, Blanchard suggests that once the agricultural activities of ploughing and lambing were finished in mid-April, the man, now miner, would make for 'the hills overlooking his farm to grub for ore ... in the shallow trenches'. Once harvest-time came, in July or August, the miner returned to full-time agricultural toil. Blanchard estimates that during his mining season, the husbandman-cum-miner might have extracted as much as two tons of ore, worth as much as £2, and as such made a significant financial gain, although this would be highly variable dependent on the type of metal involved and its contemporary market value.

Hatcher, in contrast, argues that although producing some of their own food, tinners could not provide for their entire annual needs. Full-time tinners will have had an even greater reliance on purchased food and as such lived close to or within settlements where this demand was met. In favour of Hatcher's theory, Gerrard puts forward two arguments.[100] Firstly, if tinners lived within existing settlements then subsequent development and expansion of these places would have masked or destroyed evidence pertaining to their homes. The distinguishing features of a miner's house, compared to any other, are currently unknown. Secondly, living in a settlement within easy reach of a number of tinworks would offer some protection against the failure of a single working, providing options of alternative employment. It is possible that tinners turned this to their advantage as they may have been able to choose their employment, selecting posts based on wage levels and job security.

The mapping of known streamworks and medieval settlement on Dartmoor shows no clear correlation between the two.[101] Gerrard suggests that there was little sense in establishing a settlement next to a deposit that would soon become exhausted. There is, however, a close chronological link between the expansion of settlement onto the marginal moorland and tinning between the twelfth and fourteenth centuries. Although Fox suggests that tinning may have brought both economic prosperity and people to Dartmoor, producing an increased demand for land, it is likely that, prior to the Black Death, the buoyant agricultural economy played a significant part in this expansion.[102]

There is evidence to suggest that medieval tinning may have had a considerable impact on tree cover. A pollen assemblage from Merrivale in the Walkham valley revealed a significant decrease in alder (*alnus*), and an increase in macroscopic charcoal, around Cal. AD 820–1030, and although the first documented tin-working is from the mid-twelfth century, the absence of documentation or archaeological evidence does not preclude this reduction in tree cover being linked to mining.[103] Indeed, Thorndycraft and his colleagues have linked the decline of tree pollen, mainly alder, in sequences from Okehampton Park and Houndtor from the thirteenth century onwards, with the clearance of this species from headwater floodplains exploited for tin.[104] A contemporary rise in cereal pollen suggests, however, that in part, the decline in tree cover may be due to clearance for agricultural expansion, rather than the tin industry alone. This agrees with Fox's suggestion that agricultural prosperity was a key factor in the development of the landscape at this time. In the Erme valley, a species poor pollen assemblage from a flush bog, near to the sediment study site, may be indicative of deliberate plantation, which Thorndycraft *et al.* suggest may have provided charcoal for the smelting process.[105]

The archaeology and history of tin in Devon are not necessarily always in accord, a definite chronology is lacking, and we are only beginning to understand the settlement pattern in relation to mining, but the impact on the landscape cannot be denied.[106] Within the catchment of every stream on Dartmoor there is some evidence for tin-working. The streamworks, lodeback pits, openworks, and processing sites are being steadily mapped to provide a picture which, with detailed investigation, might be assigned to the late medieval period, a visual reminder of the scale of the industry at that time.

Iron

Iron is one of the most common metals found in the country and Devon is no exception, with significant deposits along the borders of Exmoor in the north, on the Blackdown Hills in the east, on the south-eastern edge of Dartmoor, and in the limestone cliffs on the south coast around Berry Head. Indeed, sufficient iron ore could be mined to supply a small bloomery in most areas of England. Historians of the medieval iron industry have, however, focused attention on certain localities in Britain with substantial and sustained production, for example the Sussex Weald, south-west Yorkshire, and the Forest of Dean, and the working of these smaller deposits has been neglected.

A significant proportion of demand for iron was local, for agricultural and building sundries, which makes production difficult to quantify. Using Domesday records as a source, particularly those for Northamptonshire, Schubert has concluded that conflict during the reign of the early Norman kings reduced the agricultural demand for iron,[107] although local production would be quick to recover from such events. By the twelfth century there is both historical and archaeological evidence for a strong iron industry, including mines and forges, either acquired or established by monastic orders, principally by the Cistercians.[108] McDonnell has put forward a model for the technological changes in iron working at Rievaulx Abbey, Yorkshire, through to the sixteenth century, although there is evidence, from Bordesley Abbey in Warwickshire to suggest a recession in large-scale iron processing by the early fifteenth century and no recovery thereafter.[109] The purchasing of iron by the Crown for military use, either in weapons or in castle building, is well documented, with the Forest of Dean being the largest supplier.[110]

Production of iron in England is estimated to have been around 1,000 tons in 1300.[111] This was not enough to fully satisfy demand, and large amounts of iron were imported from Spain and the near continent. The availability of imported iron may have contributed to a decline in home production after 1300, although it is possible that a shortage of wood for fuel had a greater impact. Whatever the causes, there is an apparent fall in iron-making capacity prior to the advent of the Black Death. One exception appears to have been the Weald, where the proximity of London may have stimulated production. Here, the introduction of water power for bloom processing (hammering) suggests a requirement for increased capacity. Renewed demand for iron in the fifteenth century was again accompanied by the application of water power to both smelting and processing, allowing for increased production in a period of labour shortage. By the end of the century further increases in output were made possible by the introduction of the blast furnace (indirect process) in south-east England.[112] Unfortunately there is little evidence regarding the capital required for iron mining and processing, but before the fifteenth century it was probably low and came from the funds of the individual operator. With the introduction of water power to iron working processes, later evidence from the Weald would suggest that landowners were willing to invest in the fixed capital of the ironworks themselves.[113]

The techniques of mining varied from area to area depending on the nature of the iron deposits. In some areas bog iron ores, deposited by iron-rich water as a horizon in the soil, were worked at an early date by simple pitting, as at High Bishopley in Weardale, Co. Durham.[114] Stratified clay ironstone in the coal measures of south-west Yorkshire, comprising a multiplicity of thinly bedded nodular deposits, were well suited to working by means of bell pits once the outcrop deposits were exhausted. These shallow shafts, widened at their base to work as much ground as the roof stability would allow, are generally associated with coal working, although they appear to have been developed in the medieval period when iron was of greater value than coal. Abandoned workings are sometimes found back-filled with unwanted coal from the next shaft.[115] Iron deposits in the limestone occur in two forms: either as metasomatic alteration of the limestone to carbonate ironstone, as found in association with lead veins in Upper Weardale, Co. Durham, or replacement and karsitic deposits where iron oxides have replaced the limestone or filled existing cavities within the rock, as in the Forest of Dean and the Furness area of southern Cumbria. The former were worked at outcrop or in conjunction with lead/silver rake workings from the twelfth century.[116] In the Forest of Dean the irregular masses of oxide ores (haematites and limonites) were worked at outcrop by a combination of trenching and shallow tunnels, leaving a characteristic series of hollows locally referred to as 'scowles'. Similar workings might be expected in Furness. On the Weald, in south-east England, ironstone nodules, found in Cretaceous clays, were worked by pits at outcrop or by shallow shafts up to 12 metres deep.[117] In east Devon iron nodules found in the Cretaceous Upper Greensand which caps the Blackdown Hills have been worked by multiple shallow pits (Figure 2.5).[118] Six settlements scattered across the western and southern parts of Somerset that are recorded in Domesday as paying, or having paid, dues in iron blooms (the product of one firing of the bloomery furnace) probably relate to the Blackdown Hills iron industry.[119] Steeply dipping lenticular and fissure deposits (veins), as found on Exmoor and its borders (see Figure 2.6 on page 41), also appear to have been exploited at surface as linear openworks and later developed at depth by shaft mining although a positive chronology has yet to be identified.[120]

There is archaeological evidence to suggest that some iron mining

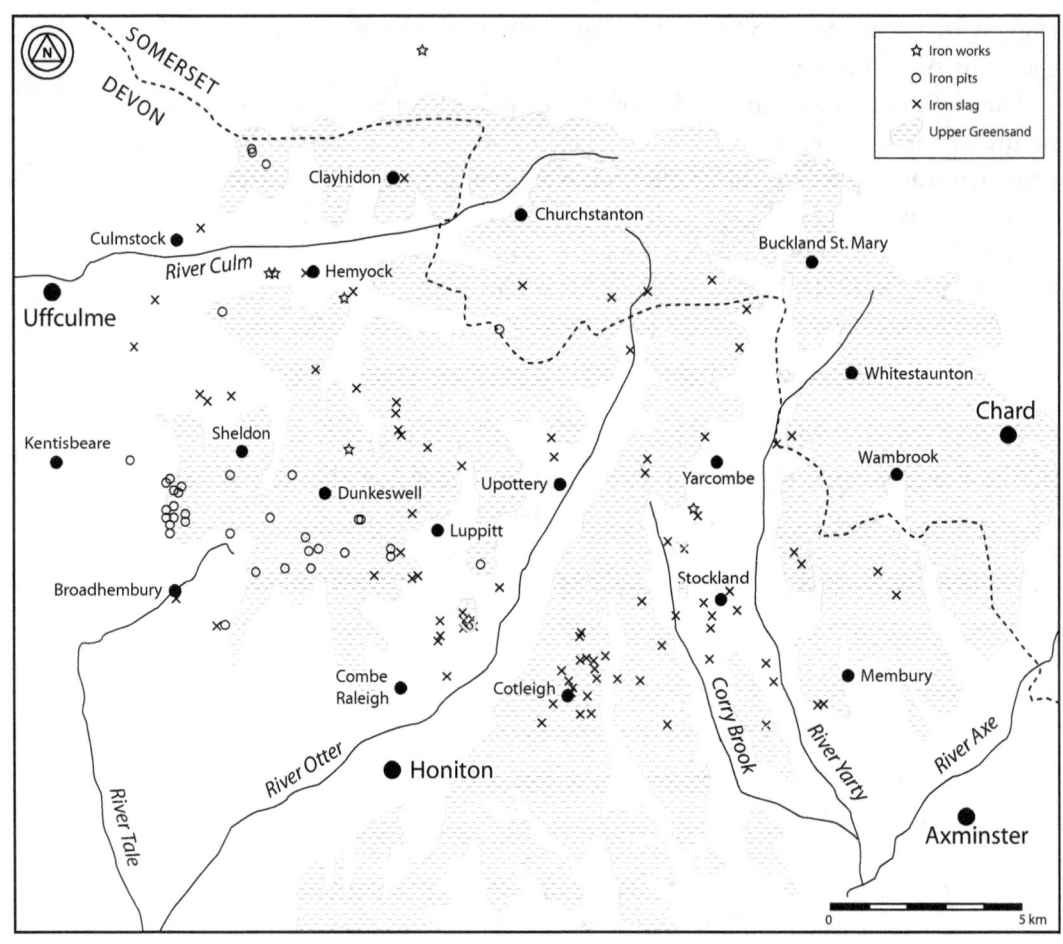

Figure 2.5: The iron-bearing Upper Greensands of the Blackdown Hills and medieval or undated iron extraction pits, iron works, and slag finds (based on the Devon and Somerset Historic Environment Record).

areas were exporting their ores for processing elsewhere. A cargo of iron ore was found in a boat, dating from the mid-thirteenth century, wrecked at Magor Pill on the Welsh side of the Severn estuary. The boat's structure was not suitable for work on the open sea and while Nayling suggested that the origin of the ore was at or near Llanharry in Glamorgan, Allen disagrees and believes that mineralogically the ore must have come from the Bristol-Mendip area: either scenario implies a regional trade in iron ores around the Severn estuary and Bristol Channel.[121] This is reinforced by further archaeological evidence from Woolaston Grange and Hill Flats, both on the banks of the river Severn, for the movement of Forest of Dean ores across the Severn to the Bristol area during the thirteenth and possibly as late as the fifteenth century, under the direct influence of the Cistercian monks of Tintern Abbey.[122] Even within an area of specialisation, iron ore might be carried some distance before being smelted, with smelting locations largely determined by the availability of fuel, wood charcoal, and, in the fifteenth century, by the availability of water as a power source.[123]

In contrast to tin, the archaeology of iron in Devon is dominated by the excellent field evidence for the smelting processes. The only archaeological dating available for the extraction of iron ores is confined to the excavation of an extraction pit at Broadhembury on the Blackdown Hills, which is probably of Roman or early post-Roman date with radiocarbon dates for the infill ranging from AD 87 to AD 550.[124] Unlike for tin, there is insufficient archaeological and historical information to understand the progression in the way that iron ores were worked, beyond an increase over time, in the depth of mining on some deposits. The iron smelting sites firmly dated to the medieval period are confined to the southern borders of Exmoor, almost certainly drawing on the siderite/hematite ores found in the Devonian rocks, and to the area surrounding the Blackdown Hills in east Devon and adjoining parts of Somerset, where there is mining of the nodular iron deposits in the Upper Greensands (see Figures 2.1 and 2.5). Medieval iron smelting in those areas shows an element of continuity from earlier, Roman/post-Roman, activity which appears to have been on a scale which satisfied more than local demand (see below). Field evidence for iron smelting is also widespread across the rest of Devon and although not archaeologically dated, it does attest to the availability of iron ores for local purposes.

Since the 1990s the iron industry in Devon has been the subject of two as yet unpublished research projects. The Blackdown Hills Ironworking Project was initiated by Devon County Council, with survey work and subsequent excavation undertaken by Exeter Archaeology.[125] This work highlighted the potential for further investigation, some of which has been realised by community projects carried out in co-operation with the University of Exeter.[126] From 1996, the Exmoor Iron Project, a collaborative project involving the University of Exeter, the Exmoor National Park Authority, the National Trust, and English Heritage has focused on an area covering west Somerset and its borders with north Devon. Like the Blackdown Hills, Exmoor's iron industry dates from the late Iron Age through to the late medieval period, but also with some evidence for continuity into the early post-medieval period.[127]

Pre-industrial iron production on Exmoor and its borders, and probably on the Blackdown Hills, appears to have been at its greatest in the Roman period. The level of ore production from Exmoor in the post-medieval period is uncertain, but modern (the late eighteenth and nineteenth centuries are known as the 'modern' period in mining history terms) production peaked in the nineteenth century at over 50,000 tons, in 1877. There is no evidence for nineteenth-century production from the Blackdown Hills.[128] Documentary evidence for iron production in the medieval and early post-medieval periods for both areas is rather limited, however, and most of our evidence comes from archaeological and paleoenvironmental investigation.

Although four iron workers (*iiii ferruarios*) are recorded in Domesday at North Molton, on the southern borders of Exmoor, for further evidence we must look to secondary sources from the post-medieval period.[129] Westcote, writing in the early part of the seventeenth century, noted that 'iron mines were sometimes wrought near North Molton and Molland', implying they were at that time inactive.[130] The Reverend Flexman, of North Molton parish, responding to Milles' survey of c.1747–56, suggests that pre-industrial production perhaps survived into the late seventeenth century when he stated that iron ores were 'worked so in the memory of man on Crowbarn Hill'.[131] The existence of these and other 'early' workings were confirmed when mining for iron resumed in the second part of the nineteenth century. The report of the Bampfylde Copper Mining Co. Ltd, when working the Stowford Mine (NGR SS713319) for iron in 1873, for example, states that 'we have got down in the old men's workings, where there is every appearance of their having a fine lode ... they must have taken away a large quantity of ore, judging from the old gunnisses or excavations.'[132] There is firmer evidence for metal-working, most likely of iron, at Prickslade in Luccombe parish. A coroner's roll of 1320 records the death of Thomas de Pyrkslade when a beam from his forge fell on his head.[133] Archaeological survey in Prickslade Combe identified quarries, a stone building, platforms, paths, and holloways and may relate to the settlement and industrial activity implied by the coroner's roll.[134] Cannell's detailed study of documentary sources for specific areas of Exmoor highlights the potential for identifying further locations of medieval iron working on Exmoor in areas not covered by her research.

Archaeological dating of the large number of pre-industrial workings across the southern borders and the central parts of Exmoor, as far east as the Brendon Hills (see Figure 2.1), has not been carried out. There

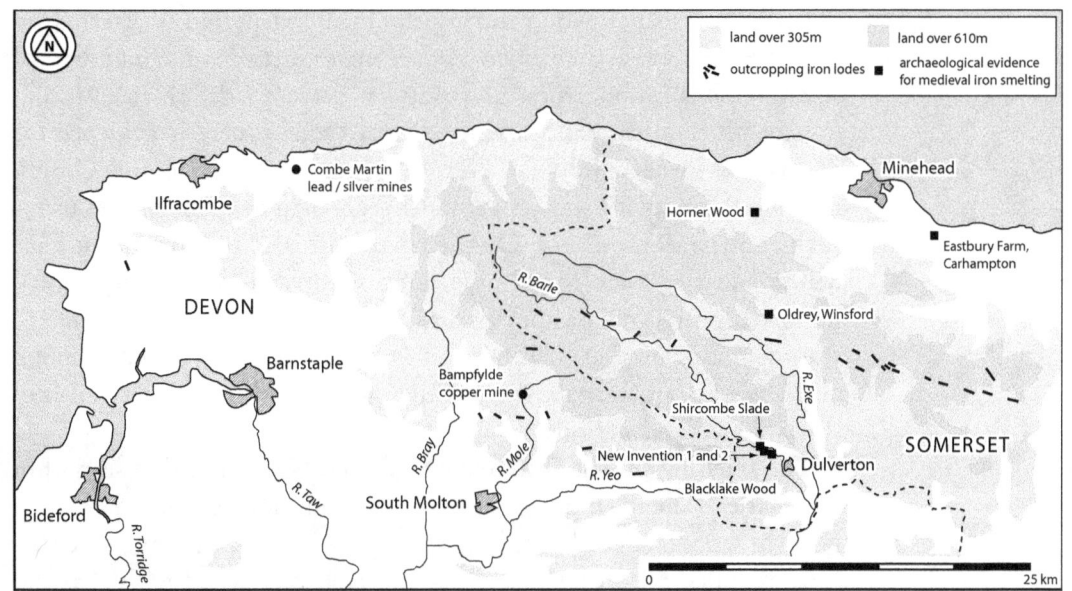

Figure 2.6: The Exmoor region showing the locations of sites with archaeological evidence for iron smelting, the copper mine at Bampfylde, and the lead/silver mines at Combe Martin.

is an argument that the vast amount of ore raised from those workings was far greater than that which could have been smelted locally in the pre-industrial periods, despite the large number of smelting sites (see Figure 2.6) and the size of the associated slag dumps,[135] yet there is no record for large quantities ore being shipped out of north Devon or west Somerset before the late eighteenth century. A patent granted to Michael Wynston, and others, in 1550 for the purposes of smelting iron using 'moor coal', that is peat charcoal, specifically refers to Exmoor as a source of raw material, but there is no evidence to suggest that any iron was produced as a result of the grant.[136] The source of the small amounts of iron exported coastwise from Barnstaple and adjoining creeks in the late sixteenth century is not known.[137] No blast furnace sites are known for north Devon, and production may be the vestiges of the bloomery (direct) process. A late, water-powered bloomery is suggested for Molland in the response to Milles' survey, where there were 'some remains of an iron furnace, a forge and mill, the latter two wrought by water' on the tenement of West Lee; but the site has yet to be identified.[138]

There is good archaeological and documentary evidence for the use of water power and the management of surrounding woodland in the early post-medieval period at Horner Wood forge on the northern slopes of Exmoor. That there was a forge at Prickslade on the edge of Horner Wood during the medieval period is attested by the coroner's report outlined above. Investigation of a number of smelting sites, radiocarbon and pottery dated to the thirteenth and fourteenth centuries, at Shircombe Slade (where there is evidence for permanent stone buildings), New Invention (two sites), and Oldrey indicate that activity was on a smaller scale than at Roman period sites such as Sherracombe Ford. On the north-east edge of Exmoor, furnace debris and slag were found associated with fifth- and sixth-century imported pottery at Eastbury Farm, in Carhampton.[139] This range of sites demonstrate that there was an iron industry operating, although not necessarily continuously, throughout the early medieval and into the late medieval periods.

The smelting sites at Shircombe Slade and New Invention sit within the Barle Valley, running north-west from Dulverton. Survey of selected woodland within this valley, upstream of the known smelting sites, revealed numerous platforms, probably associated with charcoal production.[140] The discovery of likely charcoal platforms in the Barle Valley, Horner Wood, and Culbone Wood indicates a significant charcoal industry across Exmoor.[141] Although none has been dated, it is a logical conclusion to relate at least some of these to the provision of fuel for metal processing in the medieval period. While it is clear that significant areas of Exmoor's valley woodlands have been utilised for the production of charcoal, it is perhaps wrong to assume that charcoal producers only selected wood from managed coppice during the medieval period.[142] Cannell's survey of Horner Wood revealed that the areas of coppice seen today were probably not contemporary with charcoal production during the medieval period, although that is not to deny the existence of coppiced woodland at that time.

Based on the small sizes of slag heaps, such as those at New Invention and Shircombe Slade, Bray has suggested that these waste dumps owe their creation to relatively few smelting events, and so indicates that the medieval industry was based on local provision of iron rather than long-distance export, although the export of ores remains a possibility.[143] Bray also suggests that localised iron extraction and smelting occurred across wide areas of medieval Exmoor. There is, however, as yet little evidence to relate the medieval smelting activity to the pattern of settlement on Exmoor or its borders. It is evident that the pre-industrial iron mining occupied the un-enclosed land, with an interesting relationship in areas like Ison Hill where the iron mines occupy the land adjoining the parish boundary, whereas the smelting sites are generally located in or close to the sources of fuel, the valley woodlands.

In impact of the iron industry on the wider landscape of the Blackdown Hills is rather unclear. Documentary evidence for iron working there indicates the presence of smelters (*fundator*) of Welsh origin on the demesne of the abbot of Dunkeswell in 1238,[144] but other documentary sources are restricted to occasional references to payments for iron, or manorial rents rendered in iron or blooms. Although the level of place-name evidence from the sixteenth century suggests that the iron industry was still very much in the contemporary memory at that date, there is a reliance on palaeoenvironmental evidence to gauge its significance during the medieval period.[145] In the course of investigating an enclosure and associated slag heaps at Bywood Farm, dated to the first century AD and located close to iron working of the eighth to ninth centuries, the analysis of a palaeoenvironmental sequence from a small valley mire revealed high charcoal concentrations, radiocarbon dated to the first to second centuries and seventh to eighth centuries, with a period of higher concentrations beginning in the tenth to twelfth centuries. This charcoal may be linked with the processing of iron ore in the early Roman, early medieval and into the later medieval period, and its decline in the fifteenth century.[146] Hawkins also suggests that although the lowest percentage of woodland pollen corresponds with the periods of dated iron working, since the proportions of different species vary little from the period of initial clearance in the Late Iron Age, the impact of this industry on the landscape is likely to have been relatively minor.[147] Hawkins' draws comparison with Mighall and Chambers' investigation of iron working at Bryn y Castell hillfort in Wales, where it was also suggested that the impact (of a considerably larger industry) was very localised and failed to alter the woodland composition.[148] They also suggest that fuel for the small-scale iron working would have been provided for by existing woodland, negating the need for extensive woodland management.

An overall picture for iron working in Devon is that the industry shows some continuity from a period of large-scale exploitation in the Roman period, through the early medieval, and into the late medieval period.[149] From the eleventh century onwards production levels probably increased, but they were in decline by the fifteenth although mining, if not smelting, did continue on the borders of Exmoor into the post-medieval period. The demands on woodland and its impact on the settlement pattern has not been fully assessed, but was perhaps not as significant as it was in relation to the working of tin and silver-bearing ores.

Discussion

Extensive as the resources available to the medieval miner in Devon were, the scale of their exploitation is reflective of the relative demand for specific metals and the level of knowledge at the time. Tin was undoubtedly the leading metal industry, with a product that might be exploited using undemanding technology, capable of easily satisfying fluctuations in demand, and with a significant export market. Copper had unrealised potential, restrained by the limited demands of the period; whereas iron production had an expanding local market and an, as yet un-quantified, capability of servicing external demand. Lead, as a by-product of silver production, had a ready market. Silver production itself was only restrained by the availability of resources. As will be shown in subsequent chapters it was confined by the technology available for working known resources and reliant on the expansion of knowledge to identify new silver-bearing deposits.

The landscape impact of the metal industries is equally variable. Tin-working on and around Dartmoor has left its imprint on a large scale, but, with an industry open to the involvement of a large cross-section of society and its consequent links to agriculture, it was capable of absorbing an organic growth in settlement. The iron industry, be it local or with potential to supply a wider area with both iron and its ores, made no inordinate demands on settlement nor is there evidence for the over-exploitation of woodland. Silver production, as will be examined below, made demands which left their mark on the landscape beyond the impact of the workings themselves.

Chapter 3

Silver production in medieval England and the Devon mines

DURABILITY and lack of taint made silver attractive for both decorative purposes and as prestige tableware. In the Roman period and earlier it was also used to supplement gold coin and, from the seventh century onwards, became the basis for English coinage. The first discovery of silver was probably in its native form, found at surface or close to surface in rich weathered deposits, but while silver in this form has been noted at Combe Martin there is no evidence that it made a significant contribution to overall production in Devon.[1]

At what date silver was first extracted from base metals, primarily lead, is unclear, but the cupellation process used to recover silver from metallic lead was in use in India by the third century BC and was commonly practised in the Roman period.[2] There appears to be little doubt that silver was refined from lead by cupellation on Mendip at that period as large amounts of the by-product litharge have been found on sites dated to the Roman period at Green Ore and elsewhere, but the source of the silver-bearing ores is unknown.[3] Pockets of silver-rich lead ores were known on Mendip, as is evident from the occasional reference to silver extraction during the medieval period, but it has been suggested that there was insufficient to sustain a prolonged period of mining when there were other far richer sources being exploited on the Iberian peninsula.[4] Recent reappraisal by Todd does, however, suggest that silver produced in Britain had an important regional role within the Empire, and Mendip silver may have been exploited well into the fourth century.[5] The evidence from surviving lead pigs of the Roman period suggest that, rather than being routinely recovered from all lead deposits, silver extraction was probably confined to a small number of richer deposits on Mendip and elsewhere, as in Shropshire.[6] After Britain ceased to be part of the Roman Empire the production of silver probably ceased until a renewed demand for silver in coinage from the seventh century onwards stimulated activity.[7] Increasing commercialisation and the use of coin thereafter ensured that it was always in demand.

English silver coinage

After its tentative reintroduction in the seventh century, the quality of English coinage rapidly improved. From Offa's reforms of the eighth century, through to the tenth century, that coinage was based on good quality silver. Pennies of Aethelred II's reign (978–1016), for example, averaged around 20 grains [1.3 g] of 92% silver.[8] This quality was maintained, with occasional lapses of only short duration, throughout the medieval period. The increased use of coin is marked by a growth in the number of mints, from 37 active in the mid-tenth century, growing to over 75 in the period immediately prior to the Conquest in 1066.[9] Although there were concentrations of mints close to the Channel ports and around Mendip, the spread was such that the majority of the population of England, south of the Wash, was within a short distance of a mint. The frequent re-coinages of the ninth to tenth centuries were continued by William after the Conquest and he retained a firm royal control over minting and its associated profits.[10] In the early twelfth century Henry I increased central control by reducing the number of mints as a means of combating forgery and adulteration. He also strengthened the established standard for coinage by setting a new slightly higher weight for the penny at 22 grains [1.43 g] of silver.[11] The penny was the standard English silver coin throughout the late medieval period. It was minted at a rate of 240 to the 'Tower' pound of silver [5,400 grains or 350 g] in the mid-thirteenth century, rising to 450 to the pound by 1500.[12] This rate of debasement was remarkably low when compared with continental European currencies such as the *sous or denier tournois* of France which deteriorated from 80 to the pound sterling in 1204 to 220 to the pound in 1299.[13]

Some coins of the twelfth century, for example the penny of Henry II's reign struck at the Carlisle mint in the 1160s (Figure 3.1), could have used local sources of silver and might be identified as such using techniques of lead isotope analysis.[14] The probable production levels for silver from the north Pennines in the twelfth century (below), including that from the 'mine of Carlisle', were far greater than the estimated number of coins that might have been produced at Carlisle

Figure 3.1: Silver penny of Henry II's reign struck at the Carlisle mint in the 1160s, showing the obverse and reverse faces. The north Pennine mines would have provided most, if not all, of the silver used in coins from this and other border mints of the late twelfth century.

PHOTOGRAPH: FITZWILLIAM MUSEUM, CAMBRIDGE

and other mints in the border counties.[15] If production from the north Pennines was at the levels suggested below, much of it must have been sent to other mints throughout England in the form of silver bars. Unfortunately, the records which might possibly inform us on the sources of silver used by the mints are not available until the early thirteenth century.

Regular recoinage meant that specie did not stay in use indefinitely. This form of recycling the silver in circulation, along with the requirement to exchange imported coins for the current English issues, means that most of the coins surviving in hoards or as stray finds have a mixed silver content. As the number of mints was reduced and the amount of silver entering England from the continent increased over the thirteenth century the anonymity of their content became almost universal. Greater record keeping at the mints and, from the 1290s

Figure 3.2: Silver penny of Edward I struck at the London mint in the late 1290s, showing the obverse and reverse faces. The silver could have come from Devon but it is more likely to contain a wide mix of metal from different sources.

PHOTOGRAPH: FITZWILLIAM MUSEUM, CAMBRIDGE

onwards, at the mines does mean that we have a far better idea of the impact of English silver production.[16] While the silver in a penny of Edward I's reign struck in at the London mint in the late 1290s (Figure 3.2) could have come from Devon it is more likely to contain a wide mix of metal from different sources.

English silver production in the twelfth century

The impact of silver production on the volume of English currency becomes particularly evident in the twelfth century. Prior to 1130 there is little information on the precise sources of home-produced silver. Secondary evidence such as the concentration of mints around Mendip in the tenth century and references to silver in association with areas, like Derbyshire, which were known lead producers suggest that silver was being extracted, but the quantities are unknown.[17] The *Pipe Roll* for 1130 provides the first identification of an English silver mine, the *Minierie Argi* of Carlisle.[18] This 'mine', in reality a number of separate workings, lay on the Crown demesne in the north Pennines, centred on Alston in what was to become part of the county of Cumberland. The Crown's interest in the 'mine', that portion of the produce due to it from the miners working the silver-bearing ores according to custom, was 'farmed' or leased to William and Hildret for one year for the sum of £40. It was not a new discovery as the burgesses of Carlisle had previously held the *firma* and owed the Crown 100s. (£5), but such an increase suggests that silver production was increasing and this was confirmed when Robert of Torigini noted that in 1133 the miners paid the Crown £500 in dues.[19] While the sum given should be treated with caution, it is evident that the mines were rich in silver.

Shortly after this, taking advantage of the death of Henry and the disputes which attended Stephen's accession to the English throne, David of Scotland moved his forces south and occupied the border counties of Cumberland and Northumberland along with the 'Mine of Carlisle'. David made use of the produce to strengthen the Scottish coinage and provide funds to replace those normally drawn from his English possessions.[20] With the accession of Henry II and the return of the border counties to English control we have, in the *Pipe Rolls* from 1158 onwards, a continuous record of the farm of the mines, the debts incurred by some farmers, and the amounts of lead sequestrated by the Crown.[21] Analysis of this statistical material along with far more limited evidence for the Bishop of Durham's estates in Weardale, where silver-bearing ores were also being mined, suggest that between 1130 and the end of the century the north Pennine mines produced well over two million ounces of silver, worth nearly £188,000. Such production levels had a significant impact on the amount of coin in circulation which, from a base of no more than £80,000 at the end of the eleventh century, had grown to an estimated £250,000 by 1205.[22]

With only the statistical evidence, and limited information on their organisation and management, it is difficult to assess the landscape impact of the north Pennine silver mines. Existing studies have not touched directly on the mines although some settlement studies have examined the Alston area.[23] The exact location of the workings is, however, subject to speculation although we know that the 'Mine of Carlisle' was in the liberty of Tynedale which lay in the South Tyne valley and included the manor of Alston along with large parts of what is now Northumberland to the north.[24] A substantial reduction in the farm of the mines in 1191, two years after the Bishop of Durham was granted the county and its mines, suggests that the Northumberland mines were the major producers at that period.[25] The mines on the bishop's own estates lay in the adjoining parts of upper Weardale and, possibly, in the Derwent valley south of the river. The necessary archaeological fieldwork required to identify twelfth-century silver working from the later, largely non-argentiferous, lead workings has yet to be carried out (although this is precisely what the Bere Ferrers Project has achieved: see Chapters 4–5). Until recently there has been the presumption, certainly among geologists, that the deposits in the area were, as a whole, low in silver with no real evidence for high values to support the statistical evidence. Geologists have relied on modern silver assays largely derived from ores mined at lower horizons and those do not reflect the values derived from the shallow deposits accessible in the twelfth century.[26] There is as yet no consensus on the nature of mining in the north Pennines at that period.

An element of seasonal settlement is suggested by the reference to 'shiels' or summer lodgings occupied by the miners prior to the fourteenth century.[27] Their independence and the customary regulation of mining meant that seasonal working along with dual occupation in agriculture was feasible. Transhumance, the movement of animals to upland pasture in the summer, provided the opportunity for small-scale but lucrative exploitation of shallow rich deposits which left a limited imprint in the landscape. The documentary evidence for dual occupation, where advantage was taken of opportunities in the agricultural calendar to diversify into mining, is among the earliest available for lead and tin producing areas like Derbyshire, Mendip, and the Stannaries of Devon and Cornwall in the thirteenth and fourteenth centuries.[28] It is, however, difficult to draw firm parallels with the north Pennines in the twelfth century without the supporting archaeological evidence.

The decline of silver production from the 'Mine of Carlisle' is very evident from the statistical evidence in the 1190s and the farm had been reduced to a token 10 marks (£6.66) rendered by the sheriff of Cumberland from 1212 onwards.[29] A similar trend is in evidence for the bishop's estates.[30] Thereafter the north Pennine mines were primarily lead producers. The same can be said for the mines in Yorkshire, principally those in Swaledale, which had produced some silver. Giraldus Cambrensis noted that silver was also produced near Basingwerk in north-east Wales during the latter part of the twelfth century and this may have contributed to the content of coins minted by the Welsh princes at Rhuddlan at that period, but this has yet to be tested.[31] In 1194–95 a reported silver deposit was investigated at Carreghofa, on the Welsh border north of Oswestry. Detailed preparations were made for mining and processing the silver, but less than 250 ounces were produced and the operation was a total failure.[32] Silver was also the focus of attention on Mendip prior to the fourteenth century, but there is no evidence for sustained production.[33]

Sources of silver in continental Europe

The overall picture from the end of the twelfth century is one of scarcity with little or no silver being mined in England and Wales. Silver was, however, available in increasing quantities from continental Europe, drawn into the country through an expansion of exports, particularly wool. Continental mines had provided a rich, if volatile, source of silver from the tenth century onwards. Prior to that much of northern Europe's silver supply came from mines in Asia, east of the Aral Sea, north of Afghanistan, supplemented by European production from shallow lead deposits such as the Frankish mine at Melle, near Poitiers, which supplied mints in that area until at least the late ninth century.[34]

It was central and eastern European mines which provided the bulk of the silver produced although there were significant producers elsewhere on the continent (Figure 3.3). The discovery of rich silver/

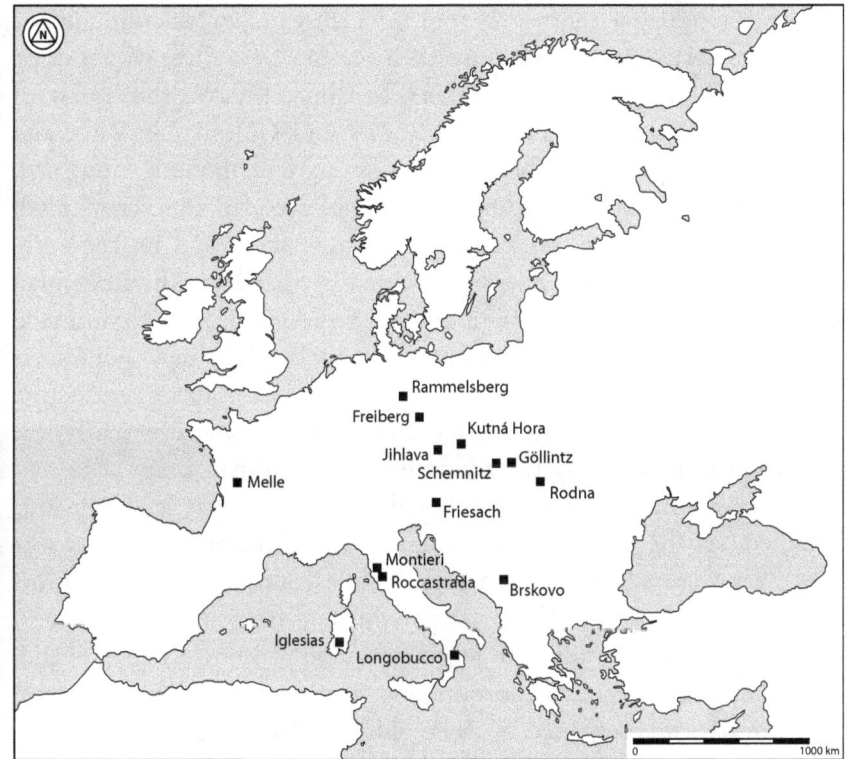

Figure 3.3:
Continental silver mines discussed in Chapter 3.

lead deposits in the Harz mountains, around Clausthal-Zellerfeld in Saxony in the late tenth century, when the mines of central Asia were in decline, established Germany as the major producer, a position it largely maintained into the post-medieval period. As mining in the Harz expanded, along with other lesser deposits opened up in the Black Forest, German silver reached every corner of the continent. Production there reached its peak in the second decade of the eleventh century, but by 1040 was in rapid decline. For the next century European output relied on residual production from the Harz and from lesser deposits, including those of northern England. In contrast to the position in England, supplied with silver from the north Pennines, the shortage of silver across continental Europe was demonstrated by a reduction in the weight and quality of coin.[35]

Relief came in the late 1160s with fresh rich deposits opened up near Freiberg, in Meissen, and, shortly afterwards, at Freisach in the eastern Alps, augmented by an expansion in production from lesser deposits at Montieri, in Tuscany. These mines were to maintain an output for nearly a century before the centres of production shifted south and east.[36] By the late thirteenth century new silver was coming from mines at Iglesias, in southern Sardinia, Jihlava, south-east of Prague, and lesser deposits at Schemnitz and Göllnitz, in Zips, and Rodna, in Transylvania (both areas now part of modern Hungary), from Brskovo, in the Balkans, and Longobucco, in southern Italy.[37] Production cycles for the richer mines were short and by the early fourteenth century outputs were eclipsed by that from the Bohemian mine at Kutná Hora, although many of the lesser mines continued in production. In Tuscany, Montieri was replaced by new, but short-lived, workings at the nearby Roccastrada mines. Balkan production reached its zenith in the mid-fifteenth century but was eventually lost to European markets by the advance of the Ottoman Empire.[38]

As central European output, based on argentiferous lead deposits, declined in the late fourteenth and early fifteenth centuries the shortages were again felt across the whole continent as many mints closed. Revival came in the latter part of the fifteenth century as new drainage technology allowed deeper working of known deposits and, more important for future production, fresh sources derived from a new ore base were opened up. The old lead based deposits now took on a new role as key suppliers to the technology developed to extract silver from argentiferous copper deposits being exploited in central Europe. Lead was an essential element in the 'saigerprocess', where it was required to draw out silver from copper. The use of argentiferous lead then had the added bonus of supplementing the silver recovered from copper.[39]

Extraction of silver from copper ores was not new, but was brought to effective, if ephemeral, commercial levels in Slovakia at the turn of the fourteenth century. As new deposits were opened up in central Europe (Bohemia and Saxony) in the mid-fifteenth century the process was to have a significant impact on European silver production. This initial central European boom could not be sustained and output quickly fell back to a base line supported by continued production from Slovakia and Thüringia (Meissen). The base line, with central European output (primarily from the mines of the Erzgebirge), rose steadily thereafter until, in the late 1520s, that from the Erzgebirge expanded rapidly to take silver production to new and unsurpassed heights.[40]

Prospecting for silver in Devon

As payments for exports to the continent drew on this vast resource the volume of coin circulating in England increased rapidly. Between £500,000 and £600,000 was in circulation by 1280 which possibly rose to over £1,000,000 by the early part of the fourteenth century before falling to half that value by the middle of the century as the English economy was drained through Edward III's military activities in northern France.[41] It was against this background that, from the mid-thirteenth century, the English Crown sought out and worked new sources of silver.

When silver-, gold-, and copper-bearing ores were allegedly discovered in a mine at 'la Hole' in North Devon in 1262 the Crown was quick to exercise a prerogative, ordering the sheriff to take possession of the workings. As yet the exact location of the mine is uncertain, although it could be at Molland where copper workings are documented on a group of tenements with 'hole' place-names.[42] Early in 1263 Crown officers were dispatched to examine the mine and payments made to them in expectation of the cost of opening up the mine. Two keepers were subsequently appointed to oversee its development, assisted by one of the king's miners, with a grant of wood for charcoal from the king's woods at Chittlehamholt. A bond of 80 marks, to be paid out of the first issues of the mine, was made to the occupier in respect of the land which he had granted to the king for the purpose of working the same.[43] There is, however, no record of any production.

There must be some doubt as to the ability of the officials, even with professional assistance, to identify and if necessary work such discoveries, as by the end of the year a group of continental miners had entered the country to prospect for mines at the king's request. They were followed, in July 1264, by Walter de Hamburg, along with seven other miners from Germany and the latter group was active at *mineris nostris cupreis, argenteis, aureis et plumbeis in comitatu Devon*. This can probably be identified with 'la Hole', although it is not mentioned by name. They were there for at least six weeks and their wages (102s.) paid out of funds held by the sheriff. Two years later Gerard de Brabaunt received £24 for 'the expenses of certain men coming from parts of Almain [Germany] to work the king's mine in Devon',[44] yet there is no further reference to these or any other mines in Devon for nearly thirty years although the Crown remained active in investigating mines elsewhere, including Ireland.[45]

The knowledge gained through the latter part of the thirteenth century could nevertheless have contributed to the decision by the Crown to work the mines at Bere Ferrers and Combe Martin, at opposite extremes of the county, from 1292 onwards.

Direct management of the Devon mines

When the silver mines of Bere Ferrers and Combe Martin were opened up in 1292 it marked a significant change in the way that English mines were operated. By choosing to manage the Devon mines directly, using its own officers and employing miners on contract, the Crown broke with the customary regulation which had governed previous silver mining activity, and continued to govern the mining of tin and non-argentiferous lead in other parts of England. Miners were recruited from the established lead mining districts across England and north-east Wales with some being pressed in to service. In Devon their skills as hard rock miners were required to work the deep seated lead/silver deposits which were outside the capabilities of the local tin miners (tinners), who were only familiar with relatively accessible alluvial deposits.

In 1294 Vincent de Hulton as keeper of the Devon mines made a return to the Exchequer detailing expenditure and production over the previous two years, subsequently confirmed in the controller's accounts submitted by William de Wymundeham.[46] These documents provide the first evidence for silver production from the Bere Ferrers and Combe Martin mines. By 1296 the latter had been abandoned as unprofitable, although they were subsequently leased to entrepreneurial interests, but Bere Ferrers proved to be a significant producer and remained under direct management until after 1349 and the advent of the Black Death. The detailed accounts that survive from this period of direct control provide a remarkable insight into the mining operation (Table 3.1).

The organisational arrangements used in the Devon mines, including the mustering of large numbers of workers, had precedents in the castle building programme initiated by Edward I.[47] At the height of medieval population growth in the last years of the thirteenth century there was evidently sufficient surplus manpower to permit large scale recruitment. Some miners in the lead mining districts probably already had reduced links to their home area, and the opportunity that gave for dual occupation in agriculture, with men having Derbyshire names

Table 3.1 Account of William de Wymondham 20–25 Edw. I [1292–97] (TNA: PRO, E101/260/6, translated with the assistance of Richard Bass)

Account of the same Master W. de Wymondham of fine silver received coming from lead from the king's mines of Birland and of Combe Martin from the 27th day of April in the 20th year of the same King E to the 15th day of September in the 25th year of the same.

20th Year	The same W. renders account of £30 of fine silver by the weight of the London Exchange[1] in one piece coming from 42 feet 25 lbs of lead from the black ore[2] of Birland smelted by the bole and of 19 feet 32 lbs of lead from the produce of the mines of Combe Martin delivered to Runcino Penyk for refining by the hand of Vincent de Hulton in the 20th year as above.
21st Year	The same W. renders account of £133 6s. of fine silver by the weight of the Exchange aforesaid coming from 230 feet 32 lbs of lead from the black ore of Birland smelted by bole of 90.5 feet 28 lbs of lead from Combe Martin delivered to the same Runcino for refining by the hand of the said Vincent in the 21st year as above.
22nd Year	The same W. renders account of £211 11s. 7d. of fine silver by weight of the Exchange aforesaid coming from 405 feet of lead from the black ore of Birland smelted by bole delivered to Master Elye and his company refiners by the hand of the said Vincent in the 22nd year as above.
23rd Year	The same W. renders account £493 9s. of fine silver by Exchange weight coming from 713 feet of lead from the black ore of Birland smelted by bole and of 120.5 feet of lead smelted from the white ore[3] by furnace delivered to the same Elye and company refiners by the hand of the said Vincent in the 23rd year as above. And from the residue of the lead of the previous years extracted from slag and ashes.
24th Year	The same W. renders account of £45 of fine silver by Exchange weight coming from 72 feet of lead from the black ore of Birland smelted by bole and hutt of 54.5 feet of lead smelted from the white ore by furnace delivered to Runcino for refining by the hand of Master W. de Wymondham and Walter de Maydenestan his controller in the 24th year as above.
	The same renders account of £783 4s. 1d. of fine silver by Exchange weight coming from 245 feet of lead from the black ore smelted by furnace and of 904.25 feet of lead from the black ore smelted by bole and hutt with the blackwork[4] of the same smelted by furnace and of 326.5 feet of lead smelted from the white ore by furnace delivered to the said Master Elye and company refiners by the hand of the said W. and W. in the 24th year as above.
25th Year	The same W. renders account of £1335 14s. 10d. of fine silver by Exchange weights coming from 445.25 feet of lead coming from the black ore smelted by furnace and 1474.75 feet of lead from the black ore smelted by bole (and from) produce of blackwork smelted by furnace and from 762.5 feet of lead smelted from the white ore by furnace with blackwork and the furnace mill. To Master Elye and company refiners by the hand of the said Master W. and Thomas de Swaneseye his controller in the 25th year above written.

Sum of all the fine silver received in (sic) the 25th year aforesaid £2932 0s. 6d. by Exchange weight from 3813.25 feet and 4.5 lbs of lead smelted by bole and hutt in the time aforesaid Birland _____⁵ 2241 feet at a rate of 2d. per lb. no allowance made for loss in refining. And of 690.25 feet of lead smelted by furnace from the black ore from which should be derived £603 19s. 4½d. at the rate of 3d. per lb. without any loss in refining. And of 1383 feet 5 lbs. of lead smelted from the black ore of Combe Martin from the white ore of Birland from which should be derived £601 3s. 1½d. at a rate of 1½d. per 1lb. without any loss in refining.

Total of silver which should remain from the refining of the lead as above no allowance having been made of the loss of the said silver in refining 2100 _____⁵ from which should remain to the refiners by the rate aforesaid £2932 5s. 6d. _____ [The latter part of this membrane is damaged and only small sections can be read, giving no legible sense to the text.]

[Account of lead delivered to the refiners]

20th Year — And there were delivered to Runcino, once master of the mint there, refiner by the hand of Vincent de Hulton in the 20th year 5_____⁵ into the account of the same Vincent from the same year 42 feet and 25 lb. of lead from the black ore of Birland smelted by bole. And in the same year 7 feet 32 lb. of lead of the produce of the ore of Combe Martin. Total 61.5 feet and 22 lbs of lead.

21st Year — And to the same Runcino by the hand of the same Vincent in the 21st year 230.5 feet and 32 lbs of the produce of the black ore of Birland smelted by bole. And the same in the same year 90.5 feet and 8 lbs of lead from the produce of the ore of Combe Martin. Total 321.5 feet and 5 lbs of lead.

22nd Year — And to Mater Elye and his company by the hand of the said Vincent in the 22nd year 405 feet of lead of the produce of the black ore of Birland. Total 405 feet of lead.

23rd Year — And to the same Master Elye and company by the hand of the same Vincent in the 23rd year 713 feet of lead of the produce of the black ore of Birland apart from 180.5 feet of pure lead transferred to the hutt. And to the same by the hand of the same 120.5 feet of lead from the produce of the white ore of Birland in the same year. Total 1014 feet including 180.5 feet of pure lead from which it is not due to be answerable for the silver.

24th Year — And to the same Runcino by the hand of Master W. de Wymundham and Lord W. de Mydenestan controller of the mines in the 24th year aforesaid there is included in the roll of the details of the lead delivered to the refiner 72 feet of lead from the produce of the black ore of Birland smelted by bole and hutt. And to the same in the same year by the hand of the same 54.5 feet of lead from the produce of the white ore smelted by furnace. And to the aforesaid Master Elye and company in the same year by the hand of the same 904.25 feet from the produce of the black ore smelted by hutt and bole including the blackwork of the same smelted by furnace. And to the same in the same year by the hands of the same 245 feet of lead from the produce of the black ore smelted by furnace mill. And to the same in the same year by the hands of the

	same 326.5 feet of lead from the white ore smelted by the furnace mill. Total 1602.25 feet of lead.
25th Year	And to the same Master Elye and company refiners by the hand of the aforesaid W. and W. in the 25th year 762.5 feet of lead from the produce of the white ore of Birland smelted at the furnace with the blackwork of the same. And to the same in the same year by the hand of the same 445.25 feet of lead from the produce of the black ore smelted by the furnace mill. And to the same in the same year by the hand of the same 1474.75 feet of lead from the produce of the black ore smelted by the bole with the blackwork of the same smelted by the furnace. Total 2682.5 feet of lead.

Sum total of the lead delivered to the refiner for refining in the years above said 6087 feet and 9.5 lbs with 180.5 feet of pure lead delivered to the hutt as above from which it is not due to be accountable for the silver. (Note – that makes the true fertile lead production 5906.5 feet and 9.5 lbs. in the five years)

[Further account of the lead delivered to the refiners]

20th Year	From the produce of the mine of Birland in the 20th year of the reign of Edward by the hand of Vincent de Hulton as is listed in the account _____ 5 lbs. of lead from the black ore smelted by bole. Item in the same year 29 feet 32 lbs. of lead from the produce of the mine of Combe Martin. Total 61.5 feet and 22 lbs. of lead.
21st Year	From the produce of the mine of Birland in the 21st year in the reign of Edward 230.5 feet 12 lbs of lead from the black ore by the hand of the same as is listed in the account. Item in the same year 90.5 feet 8 lbs. of lead from the produce of the mine of Combe Martin. Total 321.5 feet 5 lbs. of lead.
22nd Year	From the produce of the mine of Birland in the 22nd year in the reign of Edward by the hand of the same Vincent as is listed in his account 405 feet of lead. Total 405 feet of lead.
23rd Year	From the produce of the mine of Birland in the 23rd year in the reign of Edward 1014 feet of lead by the hand of the same Vincent lead smelted from the said ore _____ 5 refined at hutt. And of 713 feet smelted from Birland. And 120.5 feet of lead smelted from the white ore. Total 1013 feet of lead with 180.5 feet of lead.
24th Year	Smelting of the white ore. From the smelters smelting at the furnace from the Purification of the blessed Mary (2 Feb.) in the 24th year in the reign of King Edward until the 25th day of March 103.75 feet of lead coming from the produce of the white ore. And from the same from the said 25th day until the 21st day of May viz on the morrow of Trinity in the same year 168.75 feet of lead from the white ore. And from the same smelters from the said morrow until the 12th day of July in the same year 108.5 feet of lead. Total 381 feet of lead smelted from the white ore.
	Smelting of the black ore. From the same smelters smelting at the furnace 245 feet of lead smelted from the black ore. Total 245 feet of lead smelted from the black ore.

Smelting with the bole. From the smelters smelting at the bole from the Purification of the blessed Mary (2 Feb.) in the aforesaid year until the 25th day of March 32.5 feet of lead coming from the produce of the black ore. And from the same smelters from the said 25th day until the 21st day of May viz on the morrow of Holy Trinity in the same year 191 feet. And from the same smelters from the said morrow until the 12th day of July in the same year 91.75 feet; coming from the blackwork of the bole aforesaid 63.5 feet. Total of lead smelted with the blackwork of the same 378.75 feet of lead.

Smelting with the hutt. From the smelters smelting at the hutt from the Purification of the blessed Mary (2 Feb.) in the aforesaid year until the 25th day of March 93.25 feet from the produce of the black ore. And from the same smelters from the said 25th day until the 21st day of May 140.5 feet. And from the same smelters from the said 21st day until the 12th day of July 258.25 feet of lead. And from the same smelters from the blackwork of the same hutt smelted by furnace over the same time 125.25 feet. Total of lead smelted by hutt with blackwork of the same 617.25 feet of lead.

Sum total of lead 1602.25 feet of lead.

25th Year Smelting of the white ore. From the smelters smelting at the furnace from the feast of Epiphany (6 Jan) in the 25th year in the reign of Edward until the 22nd day of March 204.75 feet of lead from the produce of the white ore. And from the same smelters from the said 22nd day until the 26th day of April 143.5 feet of ore from the same ore. And from the same smelters from the said 26th day of April until the 24th day of June viz the Nativity of St John the Baptist 190.75 feet. And from the same smelters from the said 24th day until the 15th day of September in the same year 99.25 feet from the produce of the same ore. Item 124.25 feet from blackwork of the same furnaces. Total 762.5 feet of lead.

Smelting of the black ore. From the same from the feast of Epiphany (6 Jan) in the year above mentioned until the 22nd day of March 84.25 feet from the produce of the black ore smelted by furnace. And from the same smelters from the said 26th day until the 24th day of June 62.5 feet. And from the same smelters from the said 26th day until the 24th day of June 62.5 feet. And from the same smelters from the said 24th day until the 15th day of September 159 feet. Total 445.25 feet of lead.

The bole with its smelting. From the smelters smelting at the bole from the feast of Epiphany (6 Jan) until the 22nd day of March in the same year 272 feet. And from the same smelters from the said 22nd day until the 26th day of April 111.5 feet. And from the same smelters from the said 26th day until the 24th day of June 331.5 feet. And from the same smelters from the said 24th day of June until the 15th day of September 143 feet. Item 616.75 of lead from blackwork of the said boles. Total 1474.75 feet of lead.

Sum total of lead received this year the 25th 268.5 feet.

Total of the sum totals of lead received in the whole time aforesaid 6087 feet 9.5 lbs of lead. Of which from the white ore and from Combe Martin 1374 feet 5 lbs. And 4713 feet 4.5 lbs are from the black ore of Birland of which 180.5 feet are of purified lead of which no account should be made of the silver.

Notes to Table 3.1

1. The Mint or Tower pound – equal or nearly equal to 5400 troy grains prior to 1526 (Craig, *Mint*, xv).
2. *Nigra mina* – galena (lead sulphide).
3. *Alba mina* – cerussite (lead carbonate).
4. *Nigro opere* – blackwork (residues from the initial smelting).
5. Damaged – text not fully legible.

appearing in the return of miners impressed in north-east Wales in 1296.[48] Nevertheless, once they were working in Devon any possible opportunity for dual occupation was reduced further as the immigrant miners would have had little in the way of land available to them and were soon employed full time throughout the year.

Direct management of the mines as one operation, rather than letting out portions of the deposit to individuals or groups of miners as was the custom in the Stannaries and in lead-mining districts such

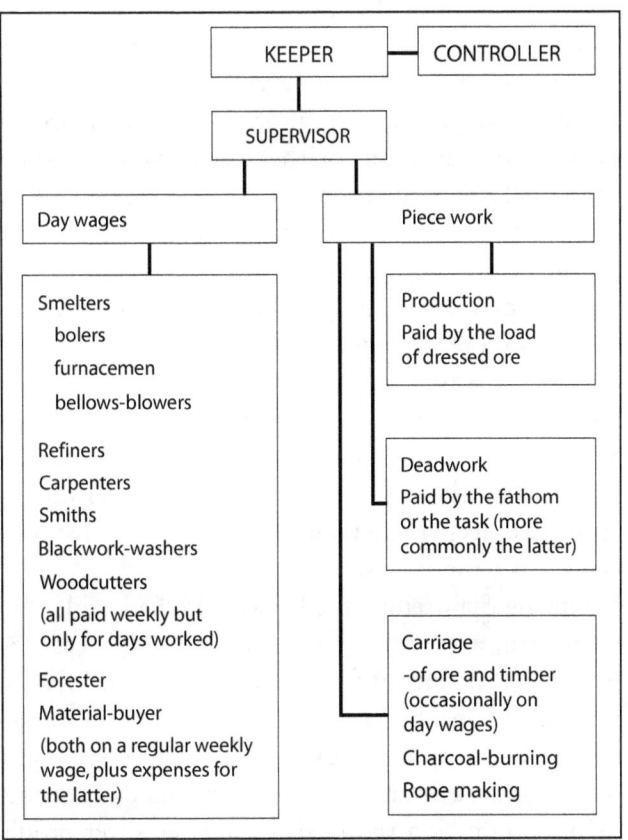

Figure 3.4: Employment structure of the Devon silver mines in the fourteenth century facilitated by direct Crown management (based on material from TNA: PRO, E101 *Exchequer Accounts*).

OPPOSITE

Figure 3.5:
Map showing the sites of known medieval mining and processing of silver-bearing ores in the Bere Ferrers region. Also shown are areas of woodland that supplied the mines with timber and fuel.

as Mendip or Derbyshire, did allow for the employment of capital intensive methods of working (Figure 3.4). This was particularly effective in organising drainage of the mines and facilitating deep working, although there is no evidence to suggest that they were the objectives for the Crown when direct management was introduced. The Crown was probably more interested in maximising the returns from the mines, following a trend established in estate management influenced by the inflation of the preceding century.

The numbers of miners employed at Bere Ferrers grew to in excess of three hundred by about 1298 in addition to one hundred tinners employed on drainage work.[49] Although the keeper admitted that the numbers were excessive, he was still considering the re-opening of the Combe Martin mines, and investigating other possible sites in Cornwall, in addition to those at Bere Ferrers. Production from the latter had grown steadily, reaching a peak of 23,228 ounces (677.50 kg) in 1297 and provided some justification for the large numbers employed. In the year to September 1297 the mines supplied nearly fifteen per cent of the silver entering the London mint.[50]

The mines at Bere Ferrers therefore proved attractive to the Crown as a means of clearing its excessive debts and to the Frescobaldi of Florence, its principal creditor, as a secure source of cash. With little evident consideration for the cost of working them, the Frescobaldi took the Devon mines on lease by an agreement dated 18 April 1299.[51] They then found the working costs so high as to deprive them of any profit, having agreed to pay 13s. 4d. per 'last' for the ore when, in their opinion it was only worth 10s.[52] Kaeuper, who has explored the role of the Frescobaldi as bankers to Edward, notes that even the sum of 13s. 4d. was in dispute, when the Crown's copy of the indenture recorded 20s. as the amount to be paid.[53] Within twelve months of the agreement with the Frescobaldi the mines were back in the hands of the Crown and Thomas de Swaneseye, controller during their tenure, was appointed keeper. He then began the task of recovering the productive capacity of the mines.

During their tenure the Frescobaldi had evidently worked for short-term gain, accounting for some 22,800 ounces (665.00 kg) of silver, but with little regard for future production.[54] Drainage adits were not maintained and had either collapsed or were in danger of collapsing, and no development work had been done to open up new reserves of ore. It took de Swaneseye four years and an expenditure of over £1,000 before production resumed. Thereafter silver production rose rapidly

Silver production

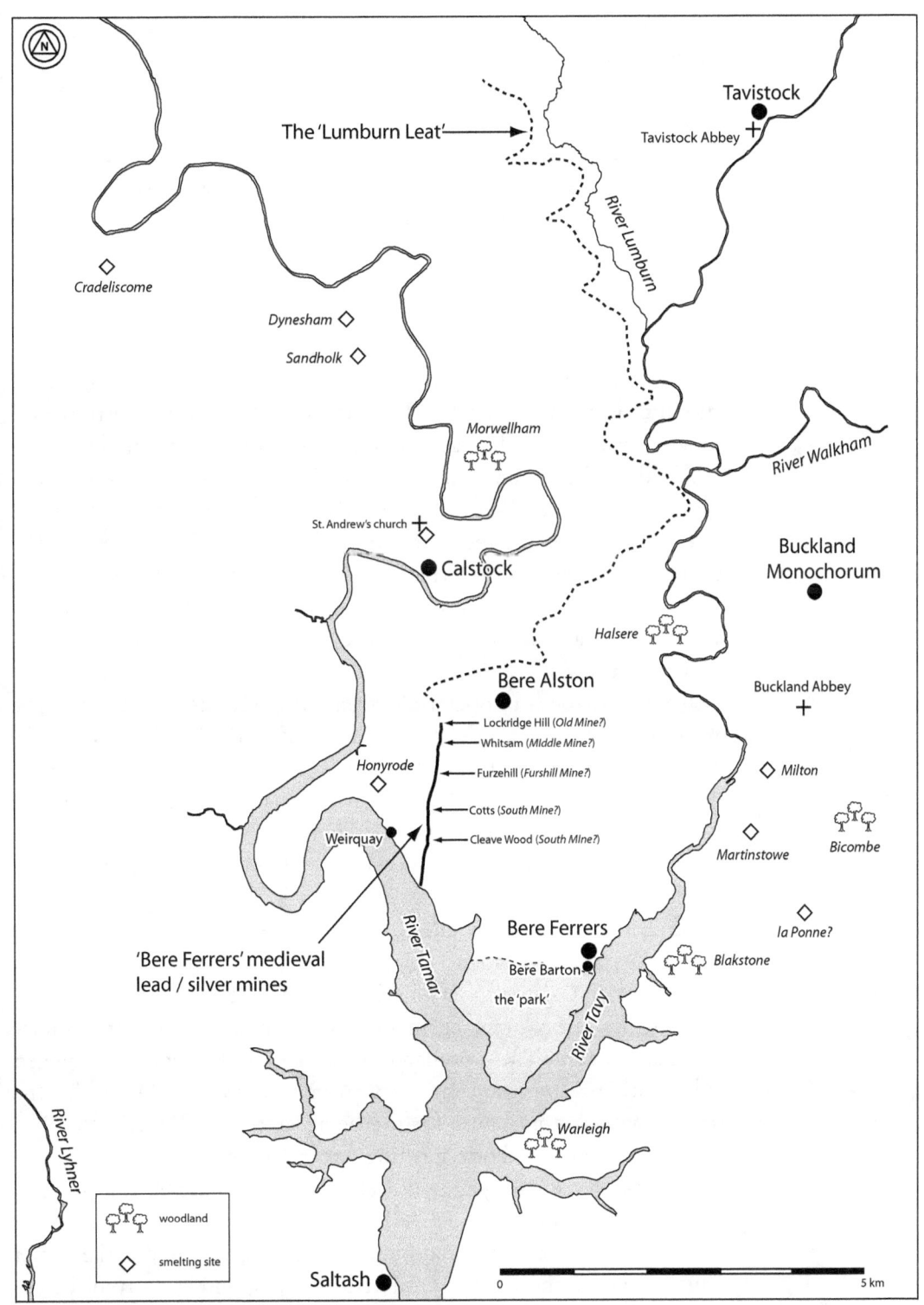

again to peak at 21,516 ounces (627.55 kg) in 1306 before falling to an average of around 8,000 ounces (233.33 kg) through to 1314.

In the first decade of the fourteenth century there were four mines being worked at Bere Ferrers on the silver-bearing deposit running from Lockridge Hill to the river Tamar south of Weirquay – the 'Old Mine', the mines of 'Middledale' (also referred to as the 'Middle Mine'), 'Furshill', and the 'South Mine' – with other workings, such as 'Hulgrove' and 'Redegrove', appearing later in the century. Field evidence suggests that these mines were active in the areas marked on modern mapping as Lockridge Hill, Whitsam, Furzehill, Cotts, and Cleave Wood (Figure 3.5). Control over operations was exercised from an administrative centre based at or close to the silver refinery. Initially, from 1292, that had been located at or near 'Martinstowe' (now Maristow) on the eastern side of the river Tavy in the parish of Buckland Monachorum. Although the mines were some distance away on the 'Birland' (Bere) peninsula, and smelting was carried out a number of sites around 'Birland' and Buckland Monachorum, the location was close to the principal source of timber on the Abbot of Buckland's lands at 'Biccombe' (Bickham) and, with the refinery on site, it was in a position to keep close control over the produce of the mines – the silver.

In 1301 the focus of operations, including refining and much of the smelting activity, was moved to Calstock after the king's woods there were assigned for the use of the mines. By Michaelmas 1302 a ferry was established 'between Birland [Bere Ferrers], where the mine is, and Calistoke where the keeper of the mines resides, boles, furnaces, refineries and the various works are'.[55] Although a specific location for the administrative centre is not given in the *Exchequer Accounts* the suggestion is that the 'curia', an enclosure containing the refinery and all the administrative facilities, may have been located close to the church at Calstock and this is considered further in Chapter 4. By 1315 the mines were again looking to Buckland for the timber required for fuel. The woodland at Calstock was evidently depleted and smelting/refining operations were moved back to Devon by the 1320s – with a store house for ore being erected at 'Martinstowe' in 1322.[56] Subsequently both smelting and refining were carried out in the water-powered 'fynyngmyll' (refining mill), in use by at least 1480, although its location is undefined.[57]

The Bere Ferrers mines continued to be worked under the direct management of the Crown until at least 1349, and the advent of the

Black Death, although at least one keeper sought greater control over the workings and their production. In 1318 the keeper, Richard de Wygorn, requested that he be granted a 'farm' (lease), of the mines at £200 per annum. The productivity of the mines had by that time fallen off, with losses of £108 in the preceding year. Nevertheless the Crown was unwilling to let them out at a fixed sum. Having checked the accounts, the view of the Exchequer was that occasional periods of loss were acceptable when 'deadwork' (development work such as sinking shafts and cutting the levels required to open up the ore deposits and allow for drainage) was underway to gain access to reserves of ore and profits would ensue. Wygorn was allowed to take the mines on farm at 12s. per load of ore raised, bearing all the costs incurred including deadwork. This was effectively a lease as Wygorn had a financial interest in the produce. The agreement was not taken beyond the three years and the following keepers appear to have reverted to wages with the Crown covering the expenses of the mines. New keepers were in place by 1320–21 and a portion of the mines was let to the Abbot of Tavistock. At his death eight years later, Wygorn was still in the king's debt.[58]

Leasing the Combe Martin mines

While the Bere Ferrers mines continued under direct management those at Combe Martin were abandoned by the Crown in 1296. In the 1320s, however, there was renewed interest and in the five years up to 1330 the mine was taken on lease by five separate adventurers including the lord of the manor, Philip de Columbariis. Prior to his being granted the mine in 1327 it was assessed by a jury who reported that its produce was then worth 10 marks having deteriorated such that 'a great part of the product of the mine is submerged and they [the miners] will be overcome by an immense abundance of water and the workmen are worried, and in this matter also they say that many veins of the same mine are totally barren'.[59]

The nature of the mineralisation at Combe Martin meant that the silver-bearing ores were present in discontinuous lenses, the 'veins' referred to in 1327, and these might be soon worked out with no chance of continuity along their strike as might be found with fault fissure deposits. The picture presented in the late 1320s is one of a mine where the readily accessible ore had been worked out. This discontinuity of the deposits, and the difficulty of predicting and prospecting for new deposits, probably contributed significantly to the lack of longevity in the working of the Combe Martin mines even into the nineteenth century.[60] By 1360, with a new discovery or improved drainage, the Combe Martin mine or mines were again being worked under a grant which covered all the mines in Devon although production levels were relatively low, totalling 9,075 ounces over five years.[61] Miners were again pressed into service, but showed a marked reluctance to stay on the mines. Twelve named miners from Derbyshire were ordered to be detained 'until they shall find security for returning to Devonshire' as they 'were set to work and abode some time at the king's wages, and have now left the works and returned to their own parts, whereby the works remain undone'.[62]

Thereafter the Combe Martin mines were worked only occasionally up to the early part of the sixteenth century, but we have no record of production levels. In the 1520s a survey of minerals in the south-west of England listed the mines as 'belonging to the King – lead holding silver',[63] and they were briefly to become a centre for the planned expansion of silver production when, in 1528, Joachim Hochstetter erected a smelting furnace there with the intention of treating ores from as far a field as Mendip.[64] The plan was short-lived and Hochstetter soon left the area. Combe Martin's most productive period was from the latter part of the sixteenth century onwards when an improved lead smelting techniques (the ore hearth process) was probably introduced. That, along with improved pumps to allow access to deeper parts of the deposit(s), allowed for the efficient working of the ores to yield significant profits, reported to be up to £20,000 but falling to £1,000 per annum after four years.[65]

At Combe Martin the impact on the contemporary landscape from mining and smelting activity in the late sixteenth century, through to the end of the seventeenth century, would certainly have been significant (Figure 3.6). The smelting/refining site was in the middle of the village, on the river Umber between the church and the manorial centre, in an area which survived as an open area, used as an animal pound, into the nineteenth century. Surrounding the site there is a deep scatter of smelting residues, primarily slag, covering an area of at least 2,500 square metres. With the subsequent building of houses and the establishment of gardens, there is little to see at surface today. Many of the early mining sites are lost under modern, eighteenth- and nineteenth-century mining to the north of Castle Street although the historic strip-field pattern is clearly modified north of Corner Lane

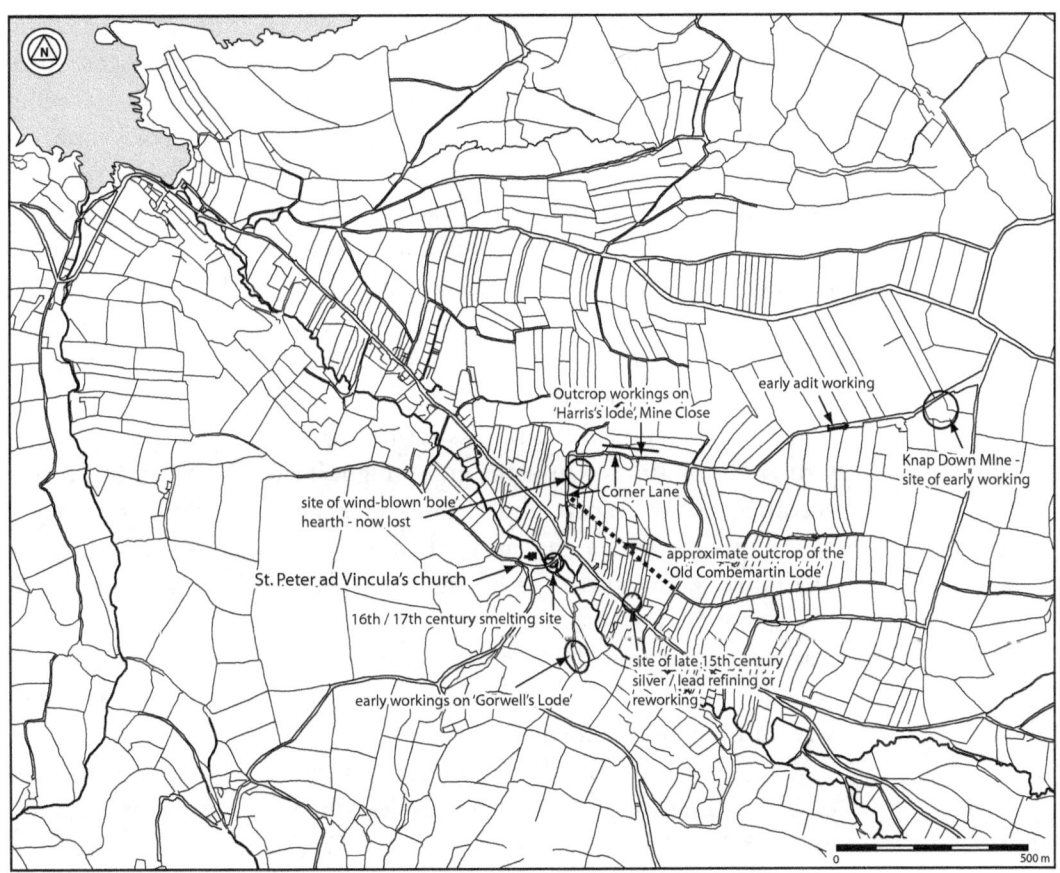

Figure 3.6: Features associated with the mining of silver-bearing ores at Combe Martin, North Devon, shown in relation to the nineteenth-century landscape (based on the Ordnance Survey First Edition Six Inch map). Intermittent mining since the thirteenth century does not appear to have influenced the layout of the strip-field system, suggesting that medieval mining was integrated into a landscape that was already settled and farmed.

to accommodate medieval or early post-medieval mining. Evidence of early smelting activity has come to light alongside Corner Lane and clearance underground on the lead/silver deposit known as Harris's Lode has revealed post-medieval workings. Above these there are earlier workings still filled with later debris and, as yet, undated.

Renewed working at Bere Ferrers

After the onset of the Black Death it was nearly one hundred years before we have recorded production from the Bere Ferrers mines. Combe Martin accounted for all the recorded production in England prior to the 1440s although Bere Ferrers was at least investigated by Crown lessees from 1360 onwards and there may have been some production for which the record has not survived. There was to be no return to direct management of the mines and from the 1350s onwards the leases granted by the Crown included both Combe Martin and the Bere Ferrers mines, a typical wording being for all 'mines of gold, silver and copper in the county of Devonshire' or, sometimes, a wider area.[66] They provided for a dead rent, if the mines were not worked, a royalty due to the Crown of from ten to twenty per cent, pre-emption on the produce to ensure the silver was delivered to the mint, and a fixed period of tenure. The lessees were expected to cover all costs and retained the profits rather than being paid wages by the Crown. From time to time the grant might include some of the powers accorded to the keepers under direct management including the right to impress miners but, as outlined above for Combe Martin, that was of limited value where there was little surplus labour in a period of falling population (although it could be said that in a period when there was a falling population, and there were agricultural tenancies to be taken up, the need to be able to impress people to work in the mines was all the more important).

Recorded production levels from Bere Ferrers were modest, at 2,440 ounces falling to an average of just less than 300 ounces per annum by 1447, but the scarcity of bullion will have concentrated effort and by 1448 production was well over 8,000 ounces per annum, and rising to over 9,000 ounces in the following three years and 5,475 ounces in the half year to 25 March 1452. During the late 1440s the Devon mines were held by the Earl of Suffolk. With his fall from favour, and subsequent murder, silver mining became briefly enmeshed in the web of intrigue which preceded a period of civil conflict in England, which has become known as the Wars of the Roses. The returns from the Devon mines during Suffolk's tenure were evidently called into question and investigations put in train to establish if silver had left the country illegally.[67] In 1449 Suffolk was licensed to export 500 fothers of lead (about 367 tonnes), the product of the Devon mines. This suggests a much reduced yield of silver, perhaps 40 oz per ton of metallic lead compared with circa 125 oz per ton in the fourteenth century; either that or some silver production was concealed.[68] Continued production into the early 1460s is suggested by the expenditure on drainage and the export of lead from the south coast ports, but no figures for silver output survive.[69] Investment in innovative pumping technology in the

1470s led to silver production in 1480/81, although the recorded level, at 1,883 ounces in that year, was disappointing.[70]

Response to the bullion crisis of the mid-fifteenth century: renewed activity at Bere Ferrers

The mercantile interest evident during Suffolk's tenure, in partnership with men from Southampton and London, was motivated by a scarcity of bullion at the period. After Suffolk's fall from favour, the king signed a grant dated 29 July 1451 making Adrian Sprynker, a German by birth, governor of the mines in Devon and Cornwall.[71] This grant recited the details of a scheme to bring the Bere Ferrers mines back into full production with increased royalties for the Crown, and this allowed Sprynker to subdivide the workings. The details of the agreements entered into by Sprynker, and confirmed by the king in *Letters Patent*, are illustrated in Figure 3.7 and allow us to identify with some certainty the areas being worked at that period. Two of the places named in the documents, 'Lockryge Hill' and 'Wittesham Down' can be clearly identified with names in use today (Lockridge Hill and Whitsam). Others, such as 'Dammewell', 'Babewell' and 'Tonnewell', are not in use today but can be identified respectively in the topography south of Lockridge Hill as the sources of the stream on the south side of the hill running west from Bere Alston towards the river Tamar, the water rising at Furzehill and running towards Whitsam, and the water rising at Little Birch and running past Gullytown to the Tamar. 'Styffe Down' is probably the high ground north of Cotts, one field on which is identified in the Tithe Award as Stone Down. The reference to 'Pechis' or Peaks Meadow is yet another link to the immigrant miners brought in from Derbyshire.

Mercantile interests are again in evidence, particularly continental European interests, among the lessees with Falron and Nicholo coming from Venice, Deloreto from Genoa, and Ram from Utrecht. Falron surrendered his lease after only a few years, having spent over £250 on drainage without profit. The other lessees appear to have had moderate success although subdivision of the mines was not continued beyond the term of the leases. Sprynker himself did not remain in post beyond a few years, perhaps a victim of factional differences in the nobility, although he did later re-appear prospecting for silver mines in the Welsh Marches on behalf of Humphrey Stafford, 1st earl of Buckingham. While the Bere Ferrers mines were subdivided,

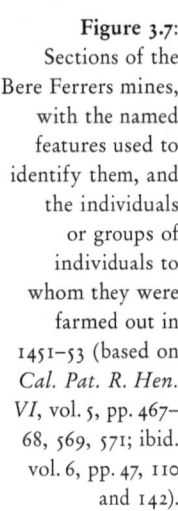

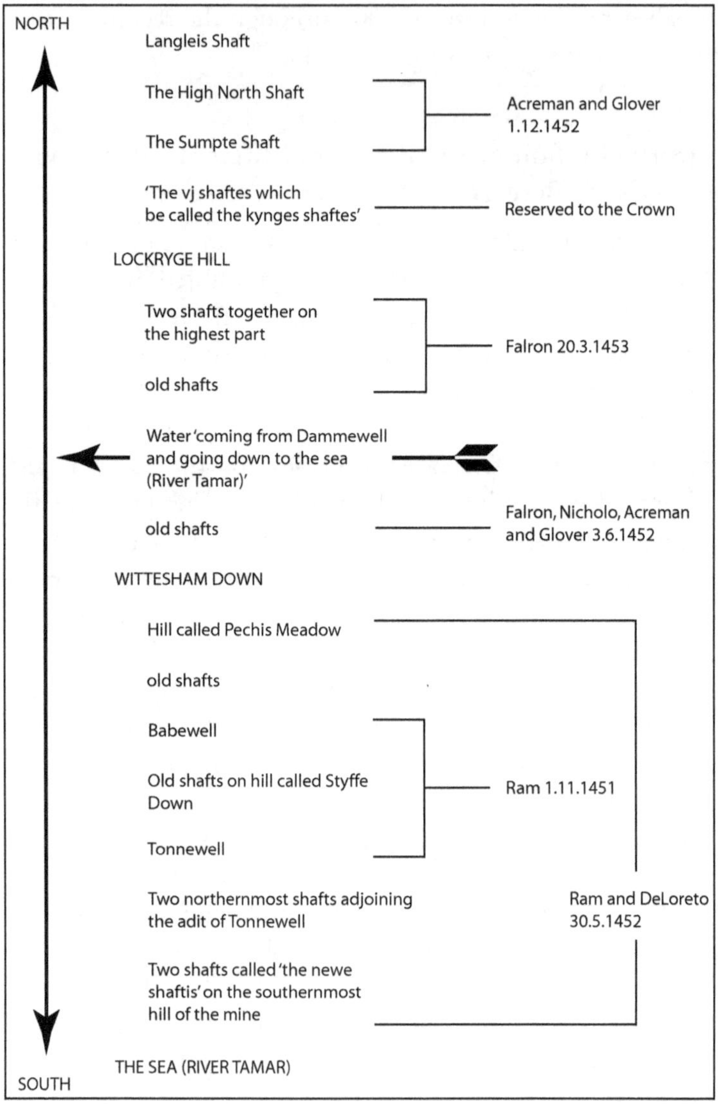

Figure 3.7: Sections of the Bere Ferrers mines, with the named features used to identify them, and the individuals or groups of individuals to whom they were farmed out in 1451–53 (based on *Cal. Pat. R. Hen. VI*, vol. 5, pp. 467–68, 569, 571; ibid. vol. 6, pp. 47, 110 and 142).

the remaining mines in Devon and Cornwall, along with six shafts reserved to the Crown at Bere Ferrers, were leased out to members of the royal household at £110 per annum.[72]

The problems of drainage at Bere Ferrers were subsequently addressed by Sir John Fogge, treasurer to the royal household, in the 1470s with the introduction of mechanised pumping. Success was, however, limited and respite from the bullion famine came not from home production but from renewed continental production. The pattern of mineralisation at Bere Ferrers meant that the deeper deposits tended to be to the south of the medieval mines, under the river Tamar, and the location for Fogge's pumps, to the north around Lockridge Hill, presented little chance for success in the long term.

Post-medieval developments

Working of the Bere Ferrers mines continued into the sixteenth century with records of expenditure up to 1538/39 but, again, there are no surviving details of production. From that period through to the latter part of the seventeenth century there is no evidence to suggest that the mines were active. In fact, the opposite appears to be the case with the parish authorities taking action to close up and make safe the old workings.[73] A new silver mine was, however, opened up at Buttspill in the late seventeenth century. This was on the same mineralised crosscourse, but well to the north of the medieval workings, and there is no evidence to suggest that it was ever worked in the earlier period. It appears on a map dated to c.1690 and is referred to as the 'New Works of the Silver Mines' in 1737.[74] It was not until the availability of effective steam-powered pumps from the late eighteenth onwards that it was possible to drain and get below the old medieval workings with mines such as South Tamar Consols producing substantial amounts of silver and lead. By that period, however, the increased value of lead was such that it was silver which was the by-product, rather than the lead as had been the case in the late medieval period.

From the 1470s production from continental European silver mines rose dramatically as mines such as those on the Rammelsberg in the Harz Mountains of what is now Germany used new pumping technology, the same technology employed at Bere Ferrers, to good effect. Although the central states once again dominated European production, new home sources were developed. The Combe Martin mines made a significant contribution to English silver production after 1580, but it was the mines of mid-Wales which had the greatest impact. The extension of English law to Wales in 1536 made the silver rich deposits, in what was north Cardiganshire, Crown property.[75]

The Crown's right to a prerogative on silver-bearing ores, along with copper and gold, was successfully tested in 1568 in the 'Case of Mines' after being challenged by Thomas Percy, earl of Northumberland.[76] Shortly before this the silver mines in Wales along with those in certain counties in England, including Devon, had been granted in perpetuity to the Society of Mines Royal. Although the Society initially worked some mines, they soon resorted to leasing them to entrepreneurial interests. Rights to silver-bearing ores in other parts of England were ceded to another monopoly company, the Mineral and Battery Works. Although the two subsequently merged, they were to lose their status when the 'Mines Royal' Acts of 1688 and 1693 removed the Crown prerogative on silver-bearing ores.[77]

It was the discovery of a rich shallow deposit of silver-bearing ore at Esgairhir, in what was north Cardiganshire, which helped precipitate the removal of Crown prerogative, but that bonanza was short-lived.[78] In reality, the easily worked shallow silver-bearing deposits in England and Wales had been worked out by the second half of the seventeenth century. Deeper working was to rely on the increased value of the lead content, rather than that of the silver. The removal of Crown prerogative along with a steady rise in price, particularly through the latter part of the eighteenth century, driven both by increased industrial demand and by the inability of long established small non-argentiferous producers, like the Derbyshire Peak, to respond did encouraged deeper working.[79] It also encouraged the re-evaluation of many silver mines including those at Bere Ferrers.

Discussion

Re-organisation of silver mining from the late thirteenth century onwards, accompanying the direct management and subsequent leasing arrangements of mines under the Crown prerogative, had facilitated deep working at Bere Ferrers and elsewhere. Private mineral owners and mining companies outside the established non-argentiferous lead mining areas from the late seventeenth century onwards had the same advantage. They were not restricted by the custom which regulated those areas, and could work or lease their mines in such a way that full use was made of technological advances in drainage and the bulk processing of the ores. The transition from customary self regulation to centralised management benefited the industry in ensuring the involvement of early capitalists and opening it to entrepreneurial interests. It put in place a form of working which was sustainable in an industrialising economy. In the latter part of the eighteenth through to the late nineteenth century, when lead/silver mining returned to Bere Ferrers, it was capable of returns unthinkable in the late medieval period.

Chapter 4

The extraction and processing of silver-bearing ores

THE EXTRACTION AND PROCESSING of silver derived from lead ores was the primary objective of the mines at Bere Ferrers from the late thirteenth through to the early sixteenth century. The value of the silver recovered at that period was far greater than that of the lead from which it was refined (Figure 4.1), and although the quantities of the latter were far greater, and it did indeed find a ready market, it should be regarded simply as a by-product where the cost and effort of mining it alone would not have been justified. This chapter will consider the documentary and archaeological evidence for the various stages in the extraction and processing of the silver-bearing lead ores, while the fuelling of this industry is described in Chapter 5.

The production of silver was carried out in essentially five stages. First there was the development work, referred to as 'deadwork' in the mine accounts, that included sinking shafts and cutting the levels required to open up the ore deposits and allow for drainage. The ore-bearing deposit itself was then mined and brought to the surface, where it was broken up and sorted (a process sometimes known as 'dressing') in order to remove the waste material (known as 'gangue'). This dressed ore was then smelted to produce a lead metal rich in silver which was then refined to recover the silver (the process known as cupellation). Both the smelting and the refining required large amounts of fuel which meant that those activities were often carried out some distance from the mines where suitable supplies of wood and charcoal could be obtained.

Sources of evidence

The documentary evidence we have for the extraction and preparation of ore is relatively limited compared with, for example, the smelting

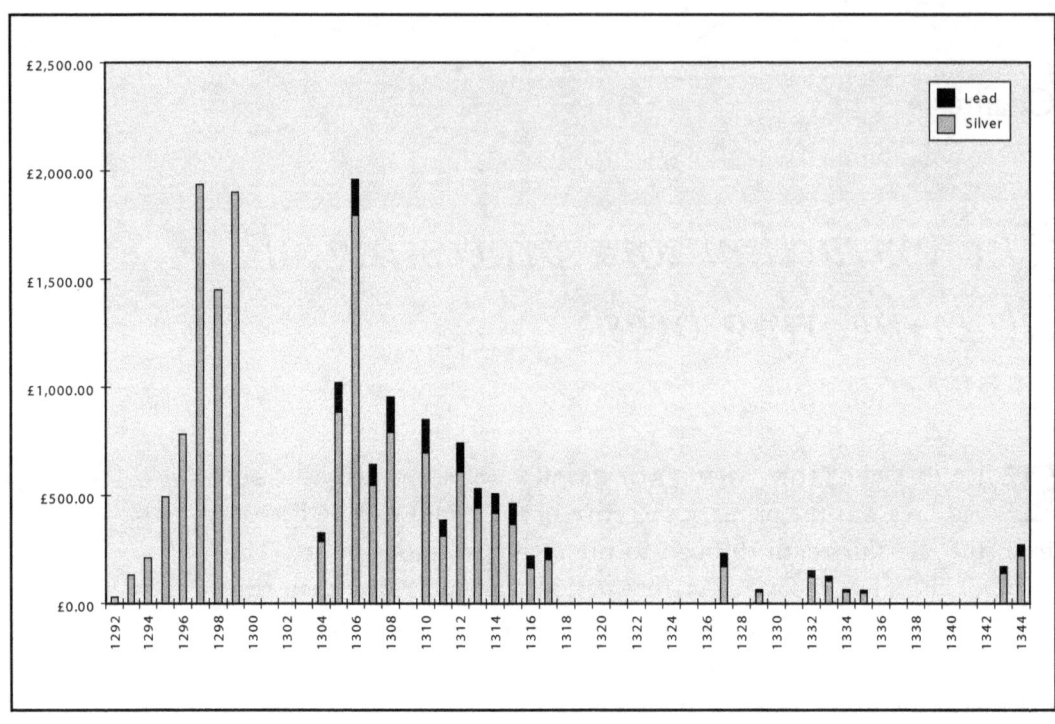

Figure 4.1: The value of silver and lead production from the Devon mines, 1292–1344 (based on material from TNA: PRO, E101 *Exchequer Accounts*). Prior to 1297 this includes production from Combe Martin, thereafter it is the product of just the Bere Ferrers mines. There is insufficient information available to determine the value of the lead produced prior to 1300.

process. This is due to the nature of the material we are working with – the financial accounts of the Exchequer – that were mostly concerned with those activities over which the mine, and its keepers, had direct control and supervision, most notably the surface operations, like smelting and refining (Table 4.1). Supervision and control of miners working on the extraction of ore and development work in scattered, relatively inaccessible, underground locations was more difficult, and so the miners were allowed a high degree of autonomy which was to be a feature of employment across the mining industry right through to the modern period.[1] Miners were paid on results, either by the amount of ore they produced or by the distance they had cut, or cleared, in the course of development work to open-up ore deposits and drive drainage adits. Generally, for ore extraction, all that is recorded in the accounts is the lump sum paid for the year. More information might be available for development work, the distance cut and the names of those employed, but no great detail on how they carried out the work. In 1342–43, for example, a typical entry in the accounts would be 'To William Node for piercing one rock blockage to make an outfall for the water – 22s. 11d'.[2]

Table 4.1: Extracts from the wage roll for the mines at Bere Ferrers in 1342–43 (TNA: PRO, E101/263/7 – translated with the assistance of Peter Mayer)

Saturday 7th December

Wages of Philip David, supervisor of the miners, receiving 14d. per week; John Dunrig' smith, for sharpening and making iron tools, receiving 12d. per week; Robert Palke, chandler and boatman, receiving 8d. per week	2s.	10d.
For 20½ stones of rope bought at Shutton, at 14d. per stone –	23s.	11d.
For 11½ stones 1 lb of tallow bought there, at 14d. per stone –	13s.	6d.
For ½ quintal of iron brought there –	2s.	9d.
Carriage of the said rope, tallow and iron in 3 casks –		3½d.
For 1 stone of steel bought –		12d.
For 8 ells of course [*grossi*] hemp bought –		16d.
For 2 sacks made thereof –		1d.
	45s.	8½d.

Saturday 14th December

Wages [as above]	2s.	10d.
For 30 seams of moor coal bought for the smithy to make tools, at 2d. per seam	5s.	
Payment to diverse miners for working and winning 14 loads 2 dishes of white ore at 5s. per load	71s.	1d.
For carriage of the same to furnace, at 1¼ for each load		17¾d.
Wages of Robert Fille, for smelting fertile lead for 4 days		8d.
For his 3 bellows blowers		12d.
For 24 quarters of charcoal made, for the task at 2½d. for each quarter	5s.	
For the carriage of the same to the furnace		9d.
Wages of 2 men washing blackwork for 4½ days and another man doing the same for 5½ days, to each per day – 2d.	2s.	5d.
For one man carrying stones to the furnace for 1 day with 2 draught animals		3d.
	£4 10s.	5¾d.

Saturday 15th February

Wages [as above]	2s.	10d.
For 20 stone 10 lb of tallow bought, at a price of 13d per stone	22s.	5½d.
For one man washing blackwork for 4 days		8d.
To Robert Fille for smelting sterile lead for 4 days –		8d.
To his 3 bellows blowers –		12d.

For washing 10 feet of sterile lead, for the task –	5s.	3d.
To William de Liccon and his colleagues for timbering 6 fathoms in the adit –	11s.	2d.
To John Hille and his colleagues for 3 fathoms of deadwork, driving in a new shaft begun at Furshille, per fathom 2s. 6d. –	10s.	
To William atte Birch and his colleagues for 4 fathoms –	12s.	
To William Node and his colleagues for driving 4 fathoms –	16s.	
To Henry Wardelowe for letting down one rock blockage, for the task –	29s.	10d.
To Robert Soper, for ½ fathom of deadwork towards *fortstull* –	7s.	6d.
	£4 19s.	4½d.

In order to understand the mining process, we must therefore supplement the documentary evidence by making comparisons with the techniques used in other mining areas,[3] examining the surface features at Bere Ferrers, and relating them to what we know about the accessible underground features. Unfortunately there has been only limited survey underground at Bere Ferrers in recent years and it was never in this project's remit to extend its work to below ground features, although some reports on previous explorations are available.[4] Some access underground is possible at Combe Martin, but comparisons with Bere Ferrers are difficult as the form of mineralisation and the local geology there meant that the techniques employed at the two sites could be significantly different (see Chapter Two). Those differences would probably have been apparent soon after the mines were opened in the 1290s.

An important feature of this project has been its strongly interdisciplinary approach that has tried to link these documentary references with physical evidence on the ground. The mine workings, forming a near continuous linear feature from Lockridge Hill southwards to Cleave Wood for a distance of over two kilometres, are identified in the accounts of the Crown from the early fourteenth century as the Old, Middle (or Middledale), Furshill, and South mines (Figures 4.2–4.3). They are, however, overlain in many places by later working dating from the eighteenth and nineteenth centuries and so identifying the medieval elements and relating them to the documentary evidence was a challenging process (see below). The more straight-forward evidence for the late medieval water supply is described in Chapter 5.

Working techniques

Within a year of their being opened, the workings were more than mere shallow trenches along the back of the lode: they had advanced to the point where artificial lighting was required and the purchase of large numbers of tallow candles is recorded.[5] One of the techniques of lead mining in the Pennines, Mendip, and north-east Wales – all areas from which the Crown recruited its miners – was to open up rich vein deposits along their length in the form of a trench, and there is some evidence to suggest that happened at Bere Ferrers. The back of the lode to the south of the minor road, immediately south of Furzehill Mine, was until relatively recently an open trench, but was apparently filled with domestic refuse in the 1920s.[6] Other sections of the lode, notably on the northern slope of Whitsam Down and beneath the road at Gullytown have been arched over at surface using mortared stone to support later surface activity. At Bere Ferrers the country rock forming the walls of the near vertical deposit was relatively stable, allowing a narrow open-work in the form of a trench to be worked to a significant depth, possibly several metres, without support. Where the silver values in the weathered mineralisation exposed at surface justified its extraction in bulk, then open-working of that type might be expected, but much of the evidence is now lost below the waste from later working.

There is little evidence that surface working of this type was used at Combe Martin where the lenticular form of the deposits, dipping at about 45 degrees in the bedding of the soft country rock, made them inherently unstable for trench working. Early nineteenth-century reports refer to early workings and 'the ground kept abroad with numberless large pieces of timber'.[7] The geophysical survey carried out along the outcrop of the Harris's Lode deposit at Combe Martin (Figure 2.2) suggests that shallow shafts were utilised early in its exploitation rather than open trenching. The earthwork mounds formed by similar shallow shafts are in evidence at Bere Ferrers, on Lockridge Hill and in Cleave Wood (Figure 4.2), and may be related to the initial working of that section of the lode deposit, probably on a portion where the silver values were not sufficiently high at outcrop to justify their working by bulk extraction from an open trench. Establishing a firm date for the workings now in evidence on the surface at Bere Ferrers is not yet possible without further investigation. All that can be said at present is that the shallow shafts,

Figure 4.2: The well-preserved earthworks of mine workings in Cleave Wood (for location see Figure 3.5).

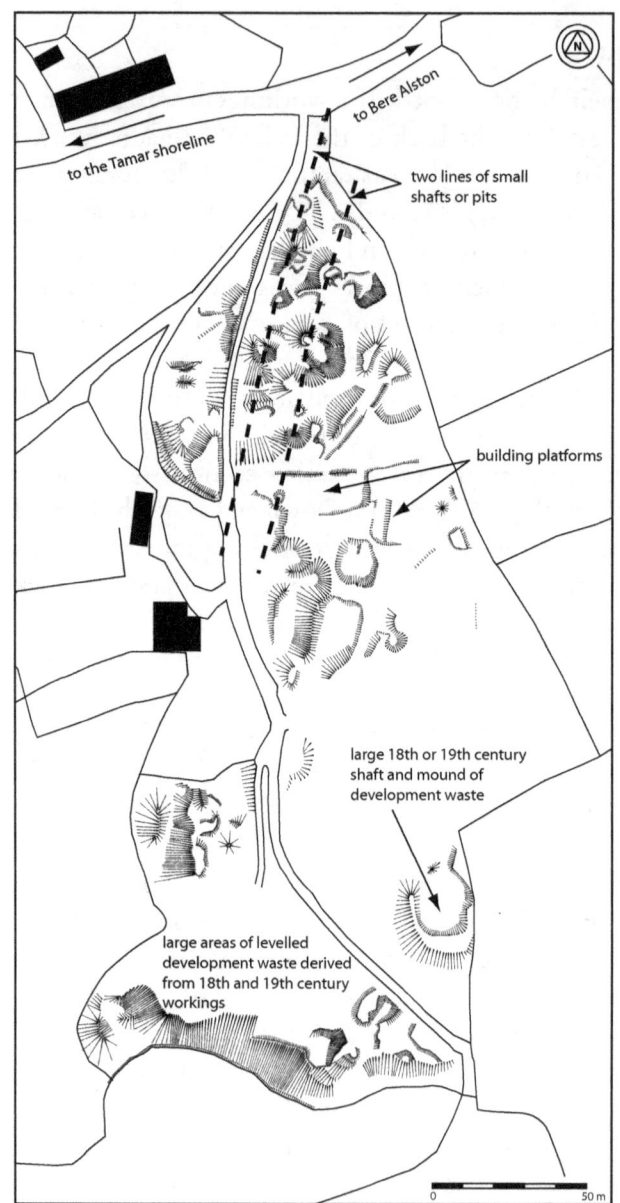

as seen in Cleave Wood, are encroached upon and partially covered by late eighteenth-century residues. They are located on the outcrop of the mineralised lode deposit, at the point of first exploitation. Given the lack of documentary evidence for working in this area between the early part of the sixteenth century and the late eighteenth-century re-working, they are probably medieval in origin.

The workings in Cleave Wood represent the southernmost point on the mineralised cross-course where there is evidence of any continuous exploitation at or close to surface. South from Cleave Wood to the river Tamar, at what was known as South Tamar Consols in the 1850s, the pattern of mineralisation takes the workable lead/silver deposits deeper in the strata, forcing mining under the river in the first half of the nineteenth century (Figure 4.3). The shallow shafts in the northern part of Cleave Wood (the eastern line of shafts in Figure 4.2) can possibly be equated to the 'newe shaftis' on the southernmost hill of the mine of 1452[8] (at the bottom of Figure 3.7).

Where open-working by trench was inappropriate, or had become impractical, shallow shafts would have been sunk on the lode. Levels, or galleries, were then opened up along the lode as far as ventilation and primitive ore handling would allow. The supply of steel tipped picks, hammers, and wedges, used by the miners for breaking both ore and rock, appears in all the accounts from the 1290s onwards and smiths were in place on all the mines to maintain them at no cost to the miners. By the first years of the fourteenth century with four separate mines and the Calstock smelting complex at work, smithing costs could amount to £37 2s., nearly 10 per cent of expenditure, in materials and labour.[9] That figure included £6 13s. 4d. in fuel costs. By the 1340s only one smith was employed, but productive working was, at that date, confined to the adjoining mines at 'Furshill' and 'Middledale'.

Fire-setting, the technique often used to assist in breaking hard rock prior to the common use of explosives, is not specifically recorded at Bere Ferrers and the lack of survey underground means that any possible physical evidence has not been identified.[10] There is some morphological evidence for its use underground at Combe Martin, although the country rock is soft enough to be worked using the pick alone, and it probably would have been used at Bere Ferrers. The wood required for the fire to heat the surface, causing entrapped moisture to vaporise, expand, and shatter the rock, would have come from the general supply to the mines and was probably not recorded separately. The only record which might have identified its use would be the provision of separate ventilation to allow the smoke to be dispersed without troubling other miners. There are references to carrying out development work specifically for ventilation, but none is linked directly to the venting of smoke from fire-setting.[11]

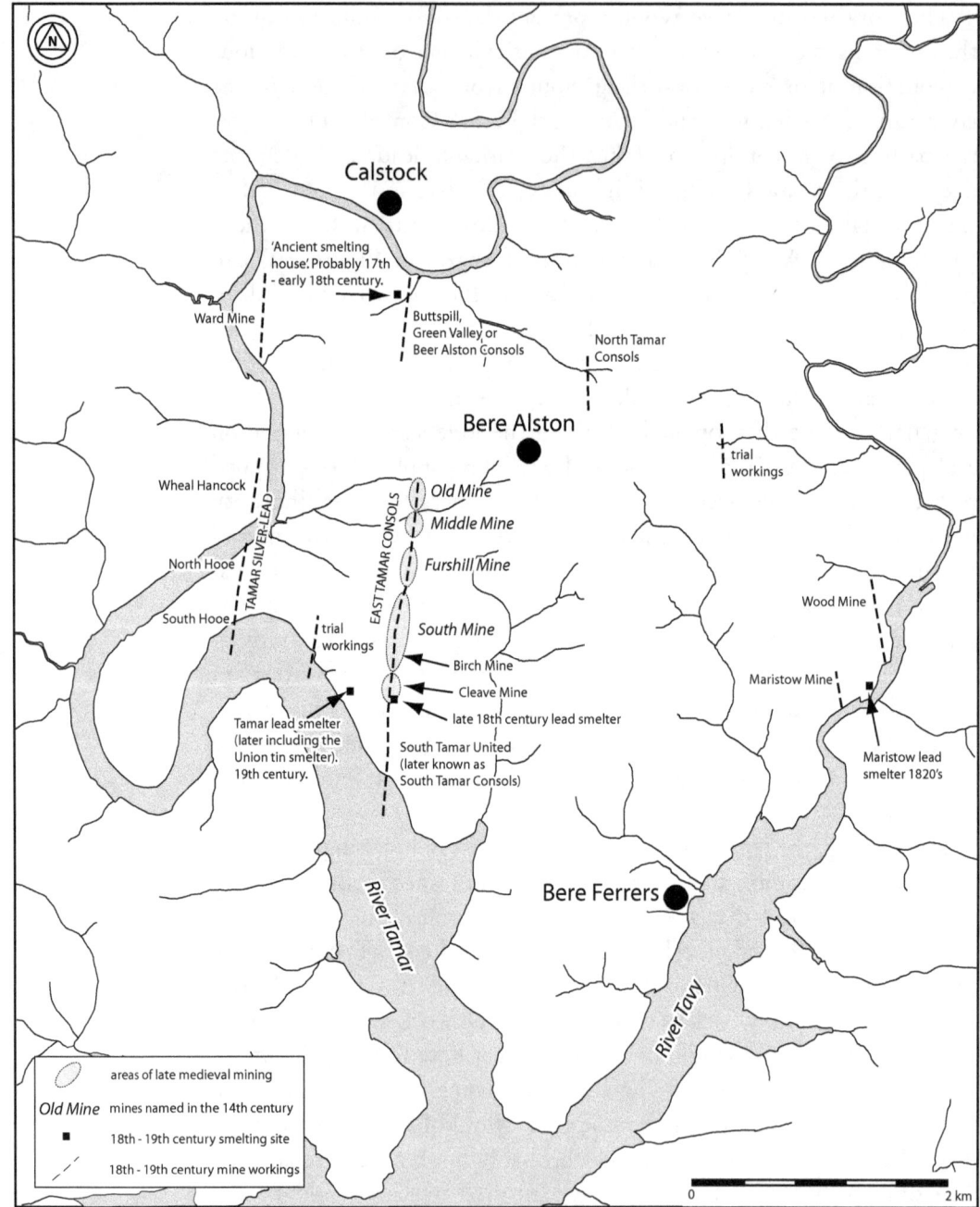

Figure 4.3: Lead and silver mines and smelters in the eighteenth and nineteenth centuries, highlighting those areas worked in the medieval period.

Drainage and the use of adits

The drainage of water entering the workings from surface, and from natural channels within the lode and associated faulting of the rock, was one of the most problematic elements of deep hard-rock mining and it was affecting work at Bere Ferrers within a few years of the mines being opened. Removal of the water could be achieved by one or a combination of three methods: free drainage to surface using horizontal galleries or 'adits', manual haulage in leather or wooden buckets, or some form of pumping. The latter method was not used until the late fifteenth century, but the first two methods were in use in the Devon mines from the beginning. By about 1297 drainage by adit was allowing the mines at Bere Ferrers to be worked through the winter months.[12] Initially the free drainage of water was probably by means of surface trenching and tinners, familiar with the drainage of relatively shallow alluvial 'stream-works', were employed specifically for the task, but it would have soon been necessary to take the drainage galleries underground to be effective.

In the early period the adits appear to have been driven either north or south along the strike of the vein in soft ground, requiring timbers for support and launders to channel water through active and abandoned workings. Such adits would be liable to collapse quite quickly and appear to have done so judging by the amounts regularly spent on their upkeep. During the Frescobaldi's tenure the adits were evidently not maintained and in the half year to Michaelmas 29 Ed. I (1301), after the mines came back under direct royal management, £56 0s. 9d. was expended in new development work and repairs, rising to £343 18s. 5½d. and £307 18s. 10d. in following years.[13] By 1308 costs were lower: £4 4s. 0d. on clearing out adits, and £10 13s. 4d. on new work. Costs fell to 32s. for repairs in 1309–10, with new work confined to 'the boring of a rock or blockage of rock from the head of the adit of the mine of Furshill lying through the middle vein, in length 60 fathoms to the south (*iacent' per mediam venam in longitudine – 60 teys versus austr'*) ... so that water could have its course to the adit' at a cost of £20.[14]

The effective use of drainage adits meant that the cost of manual water haulage could be reduced significantly. At first the cost of manual haulage was met by the miners undertaking ore extraction, although the Crown supplied the buckets and ropes required. In the five weeks to 13th August 1306, 87 new water bags were made, at 1d.

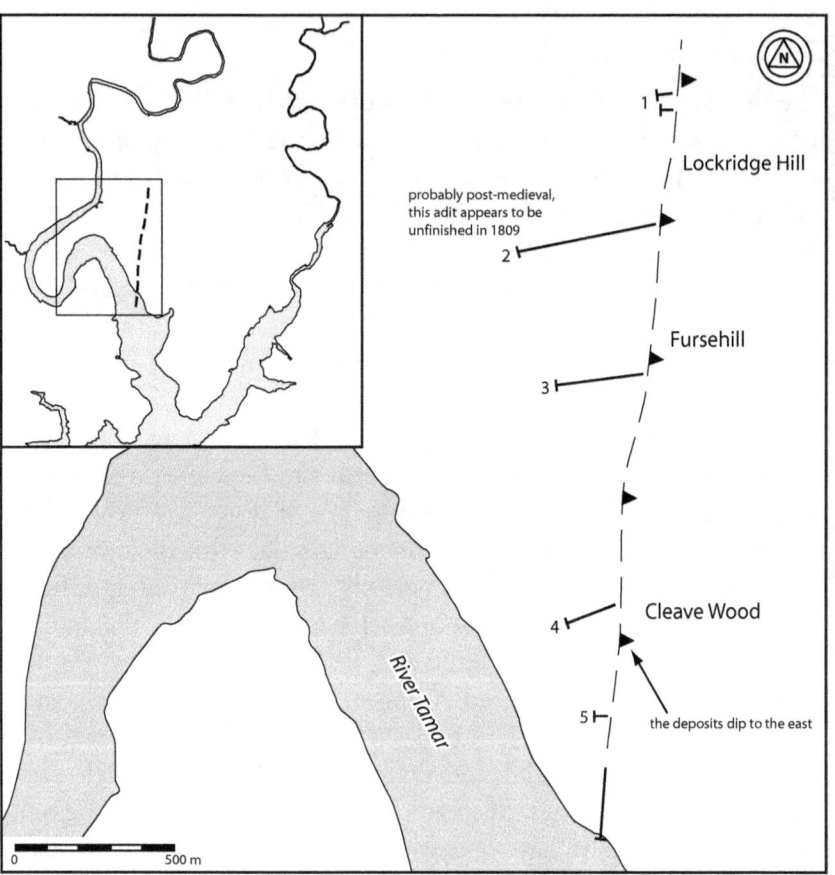

Figure 4.4: Location of the cross-cutting adits, based on mapping from the first decade of the nineteenth century (DRO, 4672A/HS/79E). Numbers 1 and 5 are undated trials at the northern and southern extremities of the lode worked in the medieval period. 2 is possibly of medieval origin, but was not completed until the nineteenth century when it was used for deep drainage of the East Tamar Consols Mine. 3 is the Furzehill crosscut probably commenced in the early fourteenth century. 4 is probably the 'adit of Tonnewell' referred to in mid-fifteenth-century documents.

each, and 31 old bags were repaired, at ½d. each; a total of 12 cow hides were then purchased at a cost of 42s. 1d. to make and repair further bags. In 1309 the wages of thirteen of the water hauliers were paid by the Crown for five months while the adit at Furzehill was extended to drain the active workings and thereafter it appears to have met the bulk of the labour costs.[15]

Unfortunately, the topography of the ground at Bere Ferrers was such that the adits driven along the lode could only have a limited effect. The lode was cut by shallow valleys to the south of Lockridge Hill and north of Cleave Wood, from which adits could be driven north or south, but in the area of Furzehill little further advantage could be gained by that technique. It was at Furzehill that there is the first reference to driving a cross-cutting adit (Figure 4.4) to increase the depth of drainage. In 1327 an adit of 50 fathoms (91.4 metres) in length was proposed to get under old workings in an area where, as

Silver-bearing ores

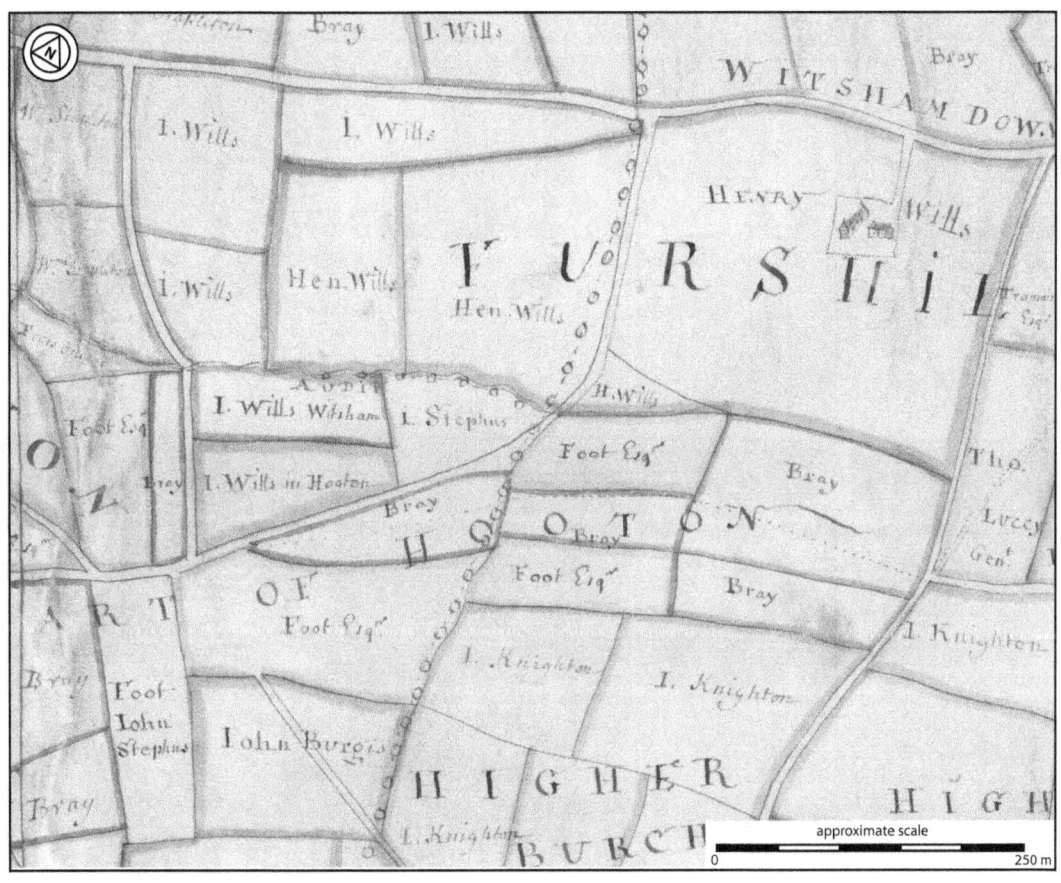

Figure 4.5: Extract from the 1737 estate map (CRO ME2424) showing the north – south oriented line of mine workings across Furzehill, and the east – west oriented, cross-cutting 'Audit' [adit] (centre, left number 3 on Figure 4.4).

stated above, there was already a long adit along the vein. The cost estimates for that new adit were correspondingly higher, £1 per fathom as against 6s. 8d. per fathom for extensions along the vein in 1309 and earlier, reflecting a slower drivage rate in hard rock.[16] It would have taken at least one and a half years to complete.[17]

It is in the Furzehill area that we have the strongest field evidence for medieval features undisturbed by later mining. A cross-cutting adit, or 'Audit' as it is named, is featured on the map of 1737 (Figure 4.5) which is some fifty years before the revival of interest in the mines and so is presumably medieval in date. The entrance, or tail, of the adit is in a shallow valley about 95 metres to the west of the lode at Furzehill where local information indicates there was a sinkage in the surface of the field in the 1950s. A shallow shaft mound in the south-western part of the woodland almost certainly marks the site of an air-shaft sunk on the back of the adit (Figure 4.6). Later mapping, from circa

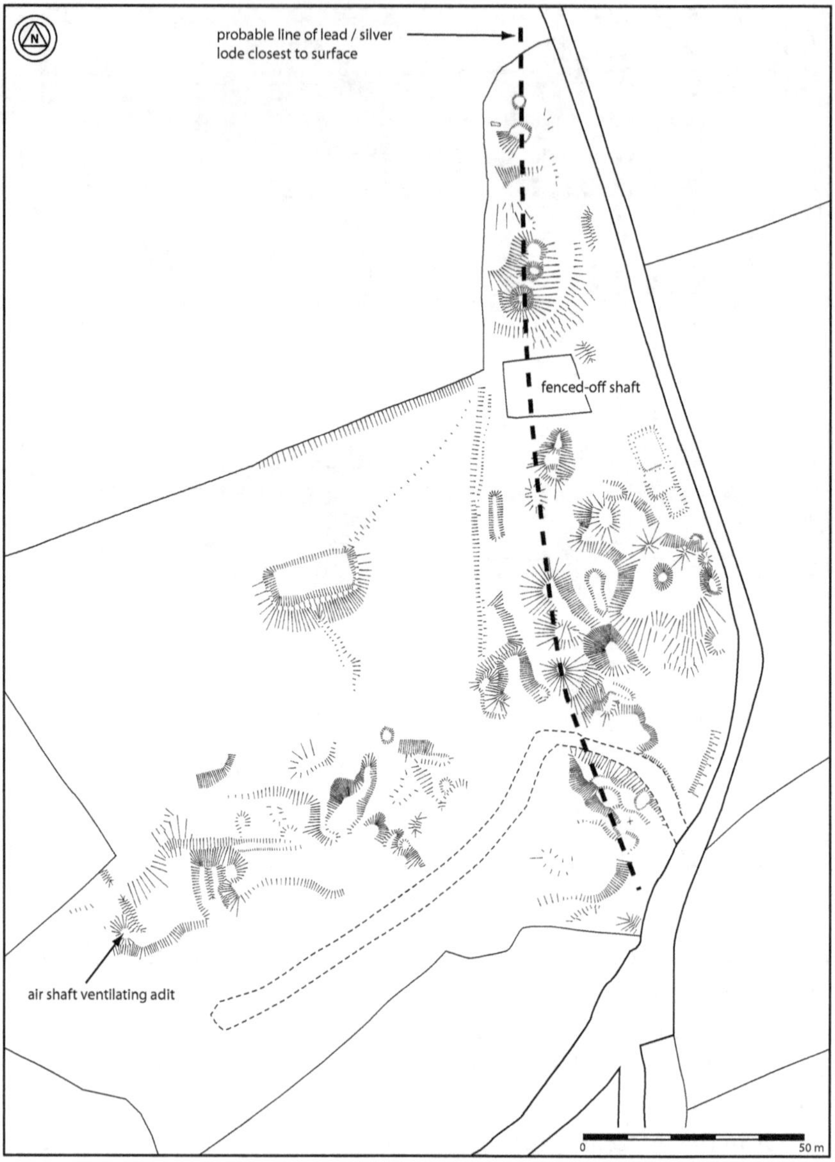

Figure 4.6:
The mining-related earthwork complex at Furzehill including the probable medieval air shaft.

probable line of lead / silver lode closest to surface

fenced-off shaft

air shaft ventilating adit

1809, does suggest that the adit tail may have been extended to the west and that it drained the lode to depth of up to 18 metres from surface which is confirmed by the results of the survey work carried by the Project.[18] It is therefore very likely that this adit is medieval in origin and possibly results from the implementation of the suggestion made to the inquiry of 1327.

The work of driving drainage adits was one of the many tasks around the mines which were not in themselves remunerative, and did not necessarily produce silver-bearing ores, but were very necessary for the continued operation of the mines. It was the commitment to such capital intensive methods on a large scale at Bere Ferrers which is a unique feature of mining in England at that period. The earliest accounts featuring such tasks, involving 'deadwork' (*mortua opera*), are for the period in the opening years of the fourteenth century when the mines were being refurbished following the departure of the Frescobaldi. Initially that involved the clearance of the old workings neglected during their tenure. It was necessary to employ miners to drain the working manually in advance of the clearance work at a cost of £1 18s. in 1300/01. Clearance and repair work, and the driving of new adits into the workings to effect adequate drainage, took until 1305. The amount of work to be carried out over those five years was planned and negotiated in advance with four groups of miners. An agreement was drawn up between the keeper (Thomas de Swaneseye), the controller and the miners' groups, specifying the total cost of £1,026 for the work to be carried out. The miners were paid at intervals through the year according to the progress made. Initially the payments were monthly, but they had become erratic by the spring of 1302.[19] It is not clear from the surviving wage rolls whether work continued throughout the year or if miners were diverted onto productive work in the summer months. There was, nevertheless, no time available for dual occupation in agriculture.

Despite this investment, even by 1312 some sections of the 'Old Mine', at Bere Ferrers, were still in poor condition. One area of workings and their associated drainage gallery had been abandoned twelve years earlier as the supporting timberwork was rotten. The gallery was cleared and the supports reinstated at a cost of £20. Similarly, a section adjacent to the 'South Mine', at *Walchemannesknot*, was cleared and the timberwork reinstated at a cost of 103s. 4d.: only when that work had been carried out was it possible to continue prospecting work in those areas. Four fathoms (7.2 metres) of new ground were cut in that year and a further 16 fathoms (28.8 metres) in the following year, at 5s. and 6s. 4d. per fathom respectively. It is clear from the amounts paid that the clearance of abandoned workings was a time consuming and probably dangerous operation, but it was necessary to allow access and drainage to new, potentially rich, ground. The untouched deposits at *Walchemannesknott* lay to the north of the old workings and drainage by manual methods was not possible until a new gallery was driven, in barren ground, from the south. Of the 12 fathoms (21.6 metres) required, half had been completed by the end of the 1313 financial year.

Three years later deadwork still continued at the 'South Mine' in the area of *Walchemannesknot*. Further north the drainage adit at 'Middledale' (the Middle Mine) was being extended into the 'Old Mine' to lower the water levels. There was already an adit draining the 'Old Mine' but workings were clearly already well below that level and reaching the limit of manual drainage. Meanwhile prospecting still continued at the level of the 'Old Mine' adit. Amounts spent on deadwork gradually rose after 1308, reaching £65 per annum in 1315, but it was necessary to tap into new deposits to sustain output which was showing signs of decline.

Once the economic limits of deep drainage using cross-cutting adits had been reached in the late fifteenth century, against a background of low population and rising real wages, innovative techniques were used in attempts to take the workings deeper below the water table. By 1480 mechanised pumping, using water-powered suction-lift pumps, was in use at Bere Ferrers, only a few years after similar systems were first used in central Europe.[20] Without an effective means of transmitting power to the pumps from a water wheel taking its supply from a remote stream it was necessary to bring water directly to the shaft head and the resulting leat, constructed between 1470 and 1480, is still a prominent feature in the landscape. The changing necessities of drainage and the introduction of new technology is described in greater detail in Chapter 5.

Development work – further prospecting at Bere Ferrers and elsewhere in south-west England

The mines at Furshill and 'Middledale' were still being expanded through prospecting work in the 1340s, but new locations at Bere Ferrers were also being explored. Speculative deadwork, boring through barren ground, was carried out next to *la Redegrove* in 1343.[21] In the following year similar work was carried out at *la Hillegrove*, *Brodebirch*, *Blakegrove*, and *Kyrkestey*. Further afield, while silver on Mendip (No. 1 in Figure 4.7) was not accessible to the Crown, being in the hands of the Bishop of Wells, prospecting was once again undertaken at the mine at *Haghelond*, in Lobbecombe near Plympton, tried unsuccessfully in 1332.[22] In Cornwall, a mine 'next St Michael's

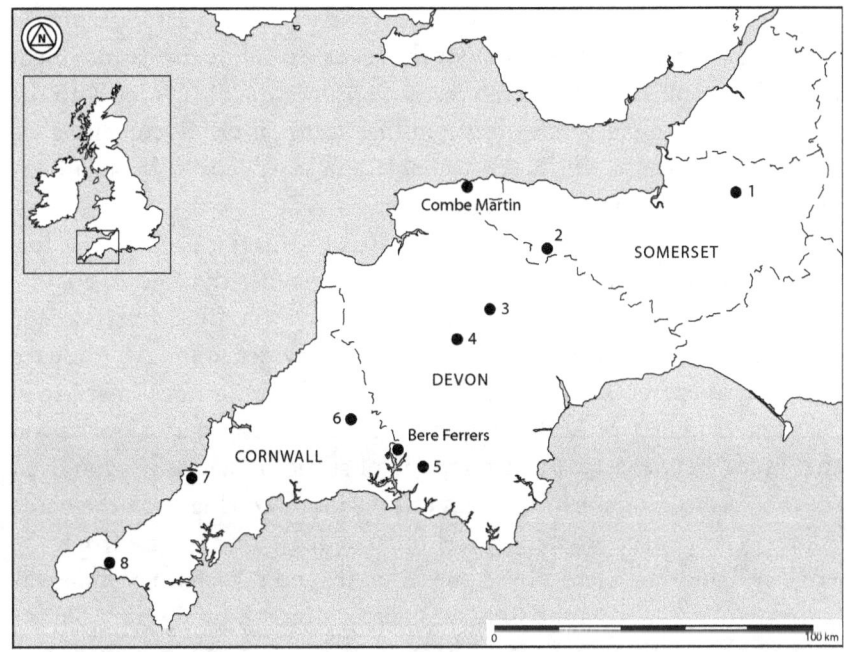

Figure 4.7:
Late medieval silver mines and trial workings in South-West England.
1: Mendip;
2: Brushford and Dulverton;
3: Brushford;
4: South Tawton and Sticklepath;
5: *Haghelond*;
6: Altarnun;
7: Cubert;
8: St Michael's Mount.

Mount' was investigated, as were other prospects in Cornwall, at Cubert in 1344 and much earlier, in 1301, at Altarnun.[23] A 'silver mine' at South Tawton and Sticklepath, in Devon to the north-east of Dartmoor, had been investigated in 1312 along with the mines at Brushford and Dulverton in Somerset.[24] Much later, in 1498, there was also a futile trial at Brushford, in Devon, in an area evidently devoid of mineralisation.[25] Although some of these prospects were worked in the post-medieval period, they all proved unproductive in the fourteenth century.

The organisation of mining

All of the miners working in the Devon silver mines were Crown employees and they were expected to work as directed by the officers of the mine. The difficulty of supervision at the workplace meant that there was an element of self regulation with payment being related to produce, which allowed for co-operation in small groups. Such groups, employed on non-productive work, are identified in the accounts by the name of an individual miner and 'his associates' without defining their relationship. The opportunity presented for a family group to raise incomes beyond that of a single wage earner is illustrated by the employment of daughters and wives on ancillary tasks associated with smelting and further processing of the residues. While the employment of family members in the extraction and processing of ores, as with the identity of all the miners working in that sector, is not documented we might expect those who were available and capable to have also been engaged in maximising family income.

For those workers recruited or pressed into service in the last few years of the thirteenth century there is no evidence that they were in a position to bring their families with them to Devon. By the first decade of the fourteenth century there is, however, evidence for the presence of family members, either immigrants or the result of local marriages. Matilda, daughter of Richard Bate who was employed as a smelter on the 'boles' between 1304 and 1317, was of sufficient age to be employed on the washing of smelting residues by 1304.[26] During 1306 the name Agnes Oppehulle, washing smelting residues, disappears from the wage roll and is replaced by Agnes Sludde, and a Richard Sludde was 'admitted to wages' as the smith at Calstock on 8th June 1310.[27] Sludde cannot be identified as a local name and Richard might be a miner, employed on extractive or other piecework prior to 1310, who probably married a local girl. She then continued to work on the mine, probably until prevented from doing so by childbirth. The evidence is scant, with only a handful of women being named in the wage rolls, and it is fragmented, but there is enough to suggest that family income could be supplemented by the employment of more than one of its members at the mines. Langdon and Masschaele have suggested that the availability of additional income could provide the incentive for early marriage, and the potential for accelerated population growth, but also contributed to low real wage rates.[28] It could indeed account for the apparent low wage rates noted by Salzman for the 1320s and 30s.[29]

There would, of course, also have been an incentive for immigrant miners to marry locally and thus establish a link to the opportunity for some form of dual occupation in agriculture even if that was confined to limited access to a small-holding. At least one family, the Smalleyes, whose origin might be identified with immigrant smelters brought in from Derbyshire in the early fourteenth century, do appear on the list of conventionary tenants in the Duchy of Cornwall's manor of Calstock in 1347 and 1356.[30]

Processing: preparing the ore for smelting

The nature of the payments to miners working on ore extraction in the Crown mines, where the miners were paid by the 'load' of ore prepared fit for smelting, means that there is very little documentary evidence available to us on the techniques used in its preparation. We must rely on what is known of the processes elsewhere, in the non-argentiferous lead fields, and comparison with the techniques used in preparing residues (slags) for re-smelting.

The ore as mined would be mixed with 'gangue' (waste material) from which it had to be separated before it could be smelted. This was achieved by hand picking and gravity separation, tasks that could be carried out by skilled youths and women, perhaps members of the miners' families if they were available. Gravity separation required the use of water and was probably carried out at dedicated sites on the streams that run close to the mines. Hand picking would have initially been carried out underground with the barren waste being packed in the old 'stopes' (the space left after ore had been extracted), and there is evidence suggesting that this was the practice at Combe Martin in the sixteenth and seventeenth centuries. The breaking of mixed material to release the ore and allow the waste to be picked out by hand and discarded was probably carried out at surface close to the lode, and a small number of special 'bucking' hammers were provided for that purpose. Although probably depicting a scene on an Italian mine, the practice is well illustrated in an illuminated manuscript of the late fifteenth century (Figure 4.8).

Identifying the sites of this ore preparation, or 'dressing', is difficult. Large amounts of waste, broken by hand, are in evidence in the valley to the south of Lockridge Hill, and smaller heaps can be found in Cleave Wood, but there is at present no firm date to indicate when that material was deposited. Hand-dressing continued in use into the modern period and it is not until the nineteenth century that we have clear evidence for the introduction of mechanised ore preparation at Bere Ferrers (a set of water-powered stamps is depicted c.1809 at Weirquay and a steam driven crusher was erected at Furshill in 1846).[31]

The use of river transport for the movement of ore to the smelters after preparation gives a general indication as to their location, but it is very general. Ore from the 'South Mine' was generally shipped out of 'Holes Hole' in the early fourteenth century whereas ore from the mines further north was carried to Lockridge Pill. Some of the named

Figure 4.8: The breaking up of ore-bearing rock allowing the waste to be picked out by hand and discarded. Although probably depicting a scene at an Italian mine, the practice is well illustrated in this illuminated manuscript of the late fifteenth century (Pliny, *Historia Naturalis*, Book XXXIII, On the property of metals: National Art Library MSL/1896/1504 – reproduced with the permission of V&A Images).

sites used for the treatment of smelting residues, *Ywille* and *la Ponne* for example, might also have been used for ore preparation, but the regular recovery of discarded smelting residues for further treatment has resulted in their total elimination as surface features. Field-walking has so far failed to find any evidence of ore preparation remote from the mining sites.

Smelting: using the wind blown 'bole' process

In the early years of the mines, up until at least the 1320s, smelting was carried out at a variety of sites scattered around the Bere peninsula and adjoining areas both east of the river Tavy and west of the river Tamar (Figure 3.5). Initially much of the ore was smelted using wind-blown hearths or 'boles' (Figure 4.9), and these would be sited in locations that took advantage of the prevailing winds channelled by the river valleys.

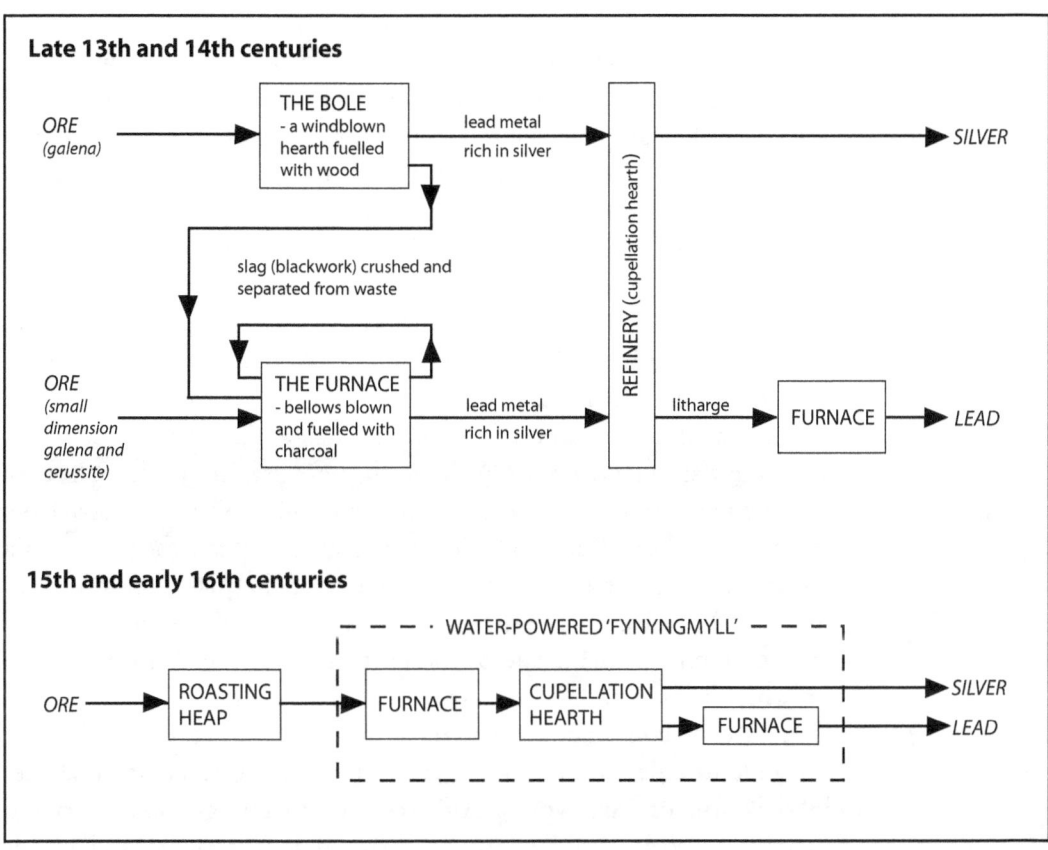

Figure 4.9: Smelting techniques at Bere Ferrers in the late medieval period. During the late thirteenth and fourteenth centuries, the return of residues, or slag, into the smelting process, after they had been crushed and the waste (with little or no metal content) had been removed, was an essential part of the technique. It introduced oxidised material into the cycle and maximised the recovery of metal from all the ore mined. The furnace in the late thirteenth and early fourteenth centuries included a number of experimental techniques. For example, the *hutt*, referred to in Table 3.1, required the return of lead into the process to combine with the silver in difficult ores – a form of liquation (see Claughton, *Silver Mining*, 166–67). By the fifteenth century the bole hearth had been discontinued and the ore was roasted to effect oxidisation before being smelted in water-powered furnaces.

Documentary evidence suggests that these 'bole' hearths were located on the high ground between Milton Combe and the river Tavy in the parish of Buckland Monachorum (the bole of 'Boclande'), immediately north of the river Tamar between Hewton and South Hooe on the Bere penisula (*Honyrode* or the bole of Birlond), and at various points on the high ground west of the Tamar in the parish of Calstock. There is, in addition, some field-name evidence for sites such as Bowl Hill near Pound in the parish of Tamerton Foliot, and, possibly, *Higesbale* which is an alternative name for Lockridge Hill.[32]

Most of the bole hearths appear to have been of substantial construction using masonry and bricks or tiles (*tegula*), although some were described as being earth structures (*bolas terreas*).[33] On some sites the bole was adapted to allow it to be turned to face the prevailing wind – the 'turnbole' – a development which appears to have been unique to the Devon silver mines. By at least 1296 a bole structure had been mounted on a moveable platform capable of being rotated about a central vertical axle to face the wind.[34] These became a permanent feature in smelting at the Devon mines. As smelting activities were moved around the 'Birland' area, and across to Calstock, in response to the availability of wood for fuel, the substantial timber base of the turnbole was uprooted for transfer to the new site. In 1303 it took seven men a whole day in 'removing the post and timber of the said bole from the ground' from Buckland Monachorum, for which they were paid 4d. each, twice the normal rate for labour at that date – heavy work indeed.[35] We have no details on the construction of the turnbole, but the development in Europe of the post-mill provided an ideal model on which it may have been based.

Re-use of construction materials and the practice of moving the turnbole to a new location appears to have left no trace of these sites in the form of earthworks. Any residues that might have survived their re-working can no longer be traced in the ploughsoil of the fields where they were probably located. The lack of slag is not surprising as it was common practice to recover smelting residues for re-smelting, and an enquiry in 1318, for example, agreed that the slag still remaining at Calstock should be sold for 100 marks on condition it was removed to a separate location for re-working.[36] Potentially, these sites may be revealed by the use of geophysical techniques, particularly magnetometry.

The bole process continued to be used up until at least the 1340s and a surprising amount of detail is available for its use at that period (Table 4.2). In the earlier years it was employed all year round, but by the 1340s suitable ore was evidently stock-piled, and smelting by this method was carried out in short campaigns every few months. The output from the mines was, by that time, very much reduced compared with the peak of production in the late thirteenth and early fourteenth centuries. In addition, the grade of ore being worked in the deeper parts of the mines would have changed. Smelting using the bole was limited to the higher grade of ore in the form of large pieces of

Table 4.2 Bole smelting costs 1342–44 (based on TNA: PRO, E101/263/8 and 10)

Date		Description			
1342	December 21st	Payment for 18 loads 3 dishes of black ore	£4	11s.	8d.
		Carriage of same to bole			23d.
		Cutting of wood for one bole plus carriage		14s.	
		To William Hacche for firing bole			15d.
		To his helpers			16d.
1343	March 1st	Payment for 8 loads of black ore		40s.	
		Carriage to the bole			10d.
		Cutting down and carrying wood to the bole		6s.	4d.
		To Jordan Kannel and 2 helpers for smelting the bole for 2 days and 2 nights			18d.
	September 20th	For cutting and carrying one bole load of wood		11s.	
		For 17 loads of black ore	£4	5s.	
		Carriage of same to bole			21¼d.
		To William Hache for the bole for a week			15d.
		To his 2 helpers			16d.
1343	November 29th	Felling and carrying all the wood for one bole		7s.	4½d.
		Payment for 13 loads black ore		65s.	
		Carriage of ore to bole			16¼d.
		To Jordan Kannel for firing the bole			15d.
		To his two helpers			16d.
1344	February 21st	For the repair of the bole house (and roofing the smithy)		3s.	1d.
	March 20th	To Phillip Purour, repairing the bole for 1 day			2d.
		For cutting wood of 1 bole load and carrying		8s.	
		For 15 loads 2 dishes of black ore		76s.	1d.
		Carriage of the same to the bole			19d.
		To Jordan Kanel for firing the bole			15d.
		To his 2 helpers for a week			16d.
	July 17th	Felling and carrying all the wood of one bole load		8s.	6d.
		Payment for 18 loads 6 dishes of black ore	£4	13s.	4d.
		Carriage of the same to the furnace and bole			23¼d.
		Jordan Kanel, for firing the bole for a week			15d.
		To his 2 helpers			16d.

September 18th	For breaking and washing the blackwork of one bole load*	4s.
September 25th	For 11 loads of black ore	55s.
	Carriage of the same to the bole	13¾d.
	For felling and carrying the wood of one bole	7s. 6d.
	Jordan Kanel for firing the bole of a week	15d.
	To his 2 helpers for the same time	16d.
	For breaking and washing the blackwork of one bole	4s.

* If compared with the rate for breaking and washing other blackwork and smelting refuse containing silver, usually 3s. per foot [70 lbs or 30.63 kg] of lead recovered, then only about one and one third feet were recoverable from the refuse of the previous firing (TNA: PRO, E101/263/5).

galena (lead sulphide), the 'black ore' referred to in Table 4.1. Anything smaller, along with the cerussite (lead carbonate), had to be smelted in bellows-blown furnaces. Those furnaces, also used to re-smelt the lead rich residues or slag (referred to as *opus nigra*, i.e. 'blackwork', in the accounts), were sited at a centralised location close to the refinery.

Before the residues, or slag, could be re-smelted they had to be crushed and washed to separate the lead-rich portions from the waste. The crushing was sometimes carried out using horse powered mill-stones, probably running on edge on a stone floor, and the washing was carried out at dedicated sites having suitable water supplies. *Ywille*, possibly a well site on the Bere peninsula close to the bole smelting site, and *la Ponne* or *Pune*, which could be on the Bere peninsula or possibly Pound at Tamerton Foliot, were used for treating the residues from the boles. They appear frequently in the accounts, but it has not been possible to identify evidence on the ground at any of the possible sites.[37]

The manual operation of bellows

The supply of water to the smelting furnaces and the refinery is addressed in Chapter 5 but through the first half of the fourteenth century there is strong documentary evidence that suggests that both relied largely, if not entirely, on the manual operation of their bellows. When the refinery was removed from 'Martinstowe' in 1301 the water wheel driving the bellows was dismantled but never re-erected at Calstock. During 1306 up to fourteen men were employed to operate the bellows at the refinery and furnaces. Only two were required at the refinery where a constant blast was not essential. The remainder were working on up to four furnaces, three men to a pair of bellows to maintain continuous operations, when a typical wage roll entry would include 'Phillip de Yal, furnaceman – 14d., three of his smelters for blowing bellows for him at the furnace – 18d.' among the payments for a week's work.[38] As Table 4.1 shows, manual operations were still in use in the 1340s and were probably only discontinued, and replaced by water power, when the decline in population led to higher wages in the fifteenth century.

Smelting and refining using bellows blown furnaces

In the 1290s the refinery and the furnaces, both of which consumed large amounts of charcoal, were located at or near 'Martinstowe' (Maristow) on the east bank of the Tavy, and the woodland at 'Biccombe' (Bickham), but in 1301 they were moved to Calstock when the woodland there was allocated to the mine. It was easier to move ore and lead to the woodland than to transport the fragile charcoal to the mines. Consequently, the furnaces returned to Maristow again when, after 1316, the wood at Calstock was exhausted and an ore store was erected there in 1322.[39] By the fifteenth century the wind-blown hearths, the boles, had been abandoned and all smelting, and refining, was carried out in a single mill, referred to as the 'fyngyngmyll'. The name implies the use of water power, considered further in Chapter 5, and a water wheel had been used to drive the refinery bellows in the 1290s.[40] It is therefore surprising, given the longevity of the Maristow site and the limitations placed on its location by the need for an adequate water supply for the mill wheel, that it has not been possible to identify it with any certainty.

The Maristow refinery and furnaces may have been sited on the stream flowing down Milton Combe perhaps using the same site as the later Lopwell Mill (NGR SX475649) but clearance and modification of that site in the late twentieth century has removed any evidence of possible earlier usage although clear earthworks for a leat do survive. Another possible site, on the north side of the river Tavy, is that used in the early nineteenth century by the lead smelter associated with the Maristow Mine. Virtually no trace of that smelter survives beyond small quantities of slag on the foreshore, some granite blocks which might have been used in its construction, and the line of the flue on the hillside high above the river. It was a reverberatory furnace, with no requirement for water power, and there is no clear evidence for any suitable leats on the hillside above and to the east of the smelter site.

While the refinery and furnaces were at Maristow, the accommodation for the Crown officers was located there and close by at 'Biccombe'.[41] When those activities were moved to Calstock they were all concentrated on the one site, the *curia*, with good access to the river but no need for a substantial water supply as all the bellows, even those for blowing the refinery hearth, were manually operated. The water wheel of the refinery at Maristow was dismantled and moved to Calstock but never re-erected there, and the only water requirement was for a limited supply brought by leat from a spring, or 'fountain', for use in washing crushed smelting residues to separate the waste before they were re-smelted.[42]

From 1301 until about 1316 the *curia* at Calstock was to be the focus of operations including the administration of the mines, refining, and much of the smelting activity. By Michaelmas 1302 a ferry was established 'between Birland, where the mine is, and Calistoke where the keeper of the mines resides, boles, furnaces, refineries and the various works are'.[43] Some idea of the relative location of the *curia* is suggested by the entry, in the wage roll for 1306, for carriage of wood 'from the water of the Tamar up to the furnaces'.[44] Earlier, in 1302, a 'ditch' or dock had been made for the boats at Calstock,[45] and the use of the river for transportation is further reflected in the reference to a boat-builder on the other side of the river (William de Clomholke/Clomhulke) [Clamoak].[46] However, instead of using the river and coastal waters to move the finished silver it was transported 250 miles overland to the exchange in the Tower of London rather than by sea. The return journey regularly took upwards of twenty days, though a quick return trip was possible in just fourteen. The consignment was, unsurprisingly, protected en-route: in 1304–05 the wage rolls record payment to an archer[47] and in 1306 Thomas Brown was paid as 'guard of the silver'.[48] In the same Wage Roll, an additional seven men were paid to travel with the 'white silver'.

There are two possible sites for this *curia*. The first is the old Rectory, 350 metres north of Calstock parish church. This has good access to the river and an adequate supply of water which could have been tapped from a resurgence feeding the stream to the north. The

Figure 4.10: Aerial view of Calstock with St Andrew's church in centre foreground, looking south across the modern town of Calstock and the river Tamar towards Buttspill in Bere Ferrers parish (copyright Steve Hartgroves, Historic Environment Service, Cornwall County Council).

ground slopes gently to the north-east, but the site of the old rectory is now a level building platform cut into the hillside which has been heavily modified over the last 150 years. An embanked area to the east does, however, appear to pre-date the building platform and is linked to earthworks forming a possible route downslope towards the river. The old Rectory was demolished in the late nineteenth century and a new Rectory (later known as Ravenscourt) built to the south, and modern residential development and stabling has obscured some of the surrounding area making it impossible to determine whether there was earlier occupation on the site.

The other possible site is in the area to the south and east of the Calstock parish church (Figure 4.10). There are occasional references to smelting and refining activity 'next to Calistok church' in the wage rolls[49] along with one reference to smelting at the *vetus castrum de Calistok*.[50] Until recently this reference to the 'Old Castle of Calstock' was assumed to be the hillfort at Castle Wood in the north of the parish, but the recent discovery that the church lies within a Roman fort (Figure 4.11) now makes this the probable site of the *vetus castrum*.

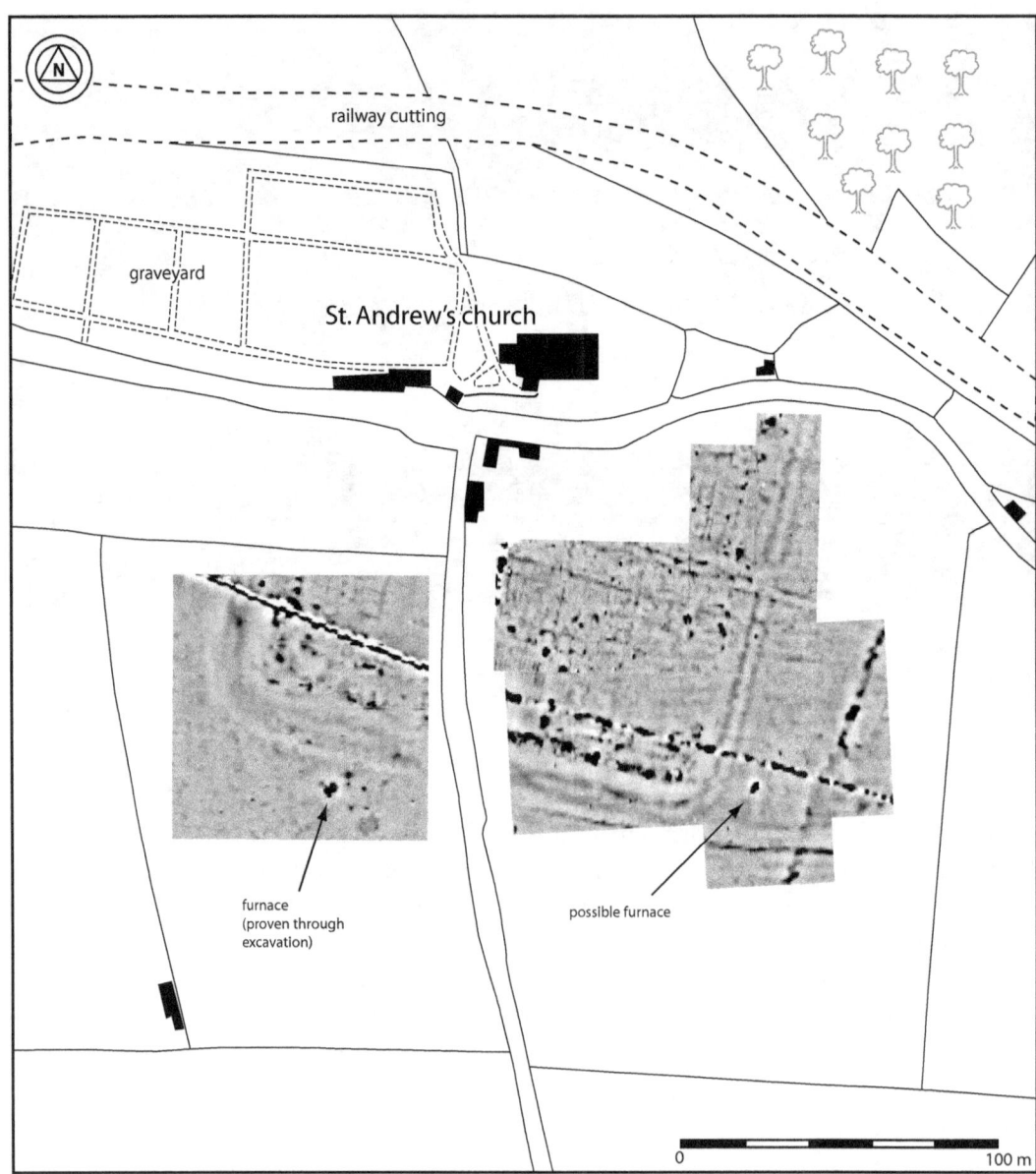

The term *curia*, suggesting an enclosure or courtyard with associated buildings, appears frequently in the wage roll for 1301/02 along with record of payments made for the relocation of two wooden buildings, a store for tallow, hides, hemp, and rope along with a small house for the use of the refiners, removed from 'Martinstowe' (Maristow). By November 1302 other wooden buildings, plastered and with thatched roofs, had been erected – a 'hall' with an upper floor, a smithy, a refinery, a store for refinery ash, and stables, with the latter being located outside and close to the gate of the *curia*. An 'engine', a horse-powered millstone used for crushing blackwork (smelting residues), had been moved there from Milton and a watercourse cut to bring water onto the site for use in washing the blackwork.[51]

In 1303/04, along with maintenance of the buildings, the keeper of the mine accounted for the costs of 'certain ditches and earthen walls' required 'to look after the lead and the whole treasury there'.[52] He also accounted for the cost of replacing the smithy which had been destroyed by fire. The refinery, 'within the curia', was plastered and re-thatched, whereas the hall was re-roofed using tiles.[53] During the following year the site was further reinforced (fortified), for the security of 'men and materials', when a ditch was dug between the *curia* and the garden,[54] and a hedge placed around the *curia*.[55] Payments were also noted in the 1306 wage roll 'for making a certain wall in the *curia* of Calistok' and the supply of stone tiles from Troghcombe. Stone tiles were again used in 1307/08 in re-roofing the refinery, after it was burned to the ground and rebuilt using wood, although other buildings, including housing for workers spending the night at the *curia*, were re-thatched.[56] When the refinery was again rebuilt in 1313/14 a mason was employed suggesting that the wooden building was replaced with a stone structure.[57] The same accounts, enrolled on the *Pipe Roll*, also note the maintenance of the hedges and ditches of the *curia* and, in the following year, the rebuilding of the controller's house, also destroyed by fire, with a tiled roof.

We therefore have a complex of buildings and smelting/refining furnaces, some within an enclosure fortified by a ditch and 'hedges'. Over a period of about fifteen years some of those building were burned to the ground, then rebuilt using wood and plaster but with stone tiled roofing. Only towards the end of the period is there a suggestion that a stone structure might have been used when rebuilding the refinery.

The refining or cupellation hearth

Cupellation, as a method of separating silver from metallic lead, has been known since antiquity and continued to be used into the twentieth century as the means of refining silver-bearing lead. Unfortunately no images survive of the process being used on mines in medieval England so we must rely on those illustrating later continental practice such as

OPPOSITE

Figure 4.11: Results of the magnetometer survey south of Calstock parish church, revealing the outline of a Roman fort, and the location of the first century AD Roman furnace.

that published by Georgius Agricola in 1557 (Figure 4.12).[58] Neither has any example of a medieval cupellation hearth been excavated in Britain, and so we must therefore rely on accounts of the process being used elsewhere[59] which can be illuminated by references in the Exchequer Accounts.

The process relied on the ability of lead and most of the impurities which might be mixed with it, small amounts of metals such as zinc and antimony, to oxidise when heated in a free supply of air whereas silver did not react and remained in its metallic form. Lead was converted to litharge (lead oxide) which, along with the oxides of the metallic impurities, might be easily separated from the silver. To achieve this silver-bearing lead metal (referred to as 'fertile lead') was heated in an enclosed hearth and subject to a blast of air from a bellows. Most of the litharge might be removed through an opening

Figure 4.12: The components of the cupellation hearth used to refine silver from lead, as illustrated by Agricola in 1556. Unfortunately no images survive of the process being used on mines in medieval England so we must rely on those illustrating later continental practice.

IMAGES SUPPLIED BY STEPHEN HENLEY

A—RECTANGULAR STONES. B—SOLE-STONE. C—AIR-HOLES. D—INTERNAL WALLS. E—DOME. F—CRUCIBLE. G—BANDS. H—BARS. I—APERTURES IN THE DOME. K—LID OF THE DOME. L—RINGS. M—PIPES. N—VALVES. O—CHAINS.

in the side of the hearth, although care was needed as the litharge was both poisonous and corrosive. To assist in the final stages of the process and the recovery of the metallic silver, the base of the hearth, or cupel, was formed from a porous heat resistant material which would absorb some of the litharge but leave the molten silver sitting on its surface. The material commonly used for the cupel in antiquity and in later, post-medieval practice was bone ash, but at the Devon mines of the medieval period ash made from oak bark, after its use in the tanning process – *curtes tannicis* or *cineres de tanno*, after it was burned – was the material of choice (see Chapter 5).[60] The cupel itself was probably formed in a wire basket or cradle, which appear occasionally in the Bere Ferrers accounts.[61] A hollow was formed in the top of the cupel in which the bead of molten silver would sit after the litharge had been absorbed by the ash.

The detailed form of the structure of the cupellation hearths used at Bere Ferrers is unknown. They would have used a lid or cover, but whether it was as substantial as that depicted by Georgius Agricola is doubtful. The hearths in Devon were not as large as that in Figure 4.12, perhaps half to two-thirds the size, and probably did not require the chains used to attach the cover to a crane; certainly no chains or crane appear in the Exchequer Accounts. The hearths used at Bere Ferrers in the first half of the fourteenth century typically processed about 550 kg of 'fertile lead' at each firing which, if compared with that illustrated by Agricola, would have required a cupel of about 0.45 metres in diameter.[62] Few examples of the cupels used for the large scale refining of silver-bearing lead have survived in the archaeological record, where they are usually referred to as 'litharge cakes', compared with the numbers of small cupels used in refining silver in jewellery or metal workshops which have been recovered from many excavation on medieval urban sites. Most large scale medieval and post-medieval cupels or 'litharge cakes' were re-smelted to recover their lead content, but some have survived on sites from the Roman period.[63]

The impact of mining on the landscape

One of the achievements of this project has been to add a spatial – landscape – dimension to the rich historical sources that survive from the medieval silver mining industry based at Bere Ferrers. In addition to the mines there were a series of processing sites for dressing and smelting the ore, and refining the silver. Due to the requirements for large amounts of fuel, the latter processes occurred in various locations some distance from the mines that led to the need for a communications system that crossed both the Tamar and the Tavy valleys. Along with the need for fuel – largely charcoal from managed woodland but also water power – this industry impacted upon a relatively extensive area.

The mine workings themselves are prominent in the landscape, from Lockridge Hill southwards to Cleave Wood, clearly cutting through an existing field system, and while individual tenements would have been divided in two this need not have caused any major disruption to agricultural practices: the mines occurred in a narrow band from just 30 to 100 metres wide that could easily have been crossed by existed roads and tracks. The landscape impact of ore processing and some ancillary activities associated with mining was even more negligible. For a brief period in the early fourteenth century the mines made their own ropes on site, rather than buying them ready made from Bridport in Dorset. The ropewalk was at an undefined site next to the river Tamar, where a house was built for the ropemakers,[64] but no evidence survives either as a place-name or as a feature in the landscape to mark its location. No ore preparation sites can be identified and neither have the smelting sites, particularly the 'bole' sites, left a lasting imprint on the landscape. At *Honyrode*, for example, on the high ground to the north-west of Hewton none of the boundaries shows any sign of respecting the medieval smelting site as is also the case on the high ground between Milton Combe and the Tavy in the parish of Buckland Monachorum. The smithing sites associated with each of the medieval mines have also left no trace: that at Calstock might be identified in the future, along with *curia*, but residues marking other sites would probably be indistinguishable from that of more modern activity.

The landscape impact of this aspect of mining, the actual extraction and processing of the ore, is therefore surprisingly limited, considering the scale of the mining activity. There were, however, other aspects of the industry – notably the requirements for large amounts of timber and fuel, and the creation of a town at Bere Alston to support the mining community – that we must now consider in Chapter 5 and 6.

Chapter 5

Fuelling the industry
The management of water and woodland

THE MEDIEVAL SILVER INDUSTRY, focused on the Bere Ferrers peninsula and surrounding parishes, was fuelled not only by manual labour and animal power, but with woodland resources and water that are the subject of this chapter. From the start of production in the late thirteenth century there appears to have been a conscious effort to preserve the long-term productivity of the region's woodland resources as they came under increasing pressure, while from the outset there was also a need for water power in the processing of ores, and later, in the fifteenth century, to drive drainage pumps. The latter, as will be seen, saw the construction of an extensive leat which drew on water from west of Tavistock, some sixteen kilometres away.

The demands on woodland

The provision of woodland was essential to a variety of medieval rural industries including iron and other metals, glass, pottery, and tile, which all required wood for fuel, and to a lesser degree the leather and soap industries which required bark and wood ash respectively.[1] In some cases the regulation of metal mining according to custom, as was the general practice prior the opening of the Crown mines in Devon, included the right to freely cut woodland for fuel and building materials, as in the north Pennines, or obliged the mineral owner to supply wood as fuel for smelting, as was the case at Grassington in Yorkshire.[2] The keepers and, after 1350, the lessees of the Crown silver mines in Devon, however, had no general right to free wood and while the supply of wood for timber and fuel might be ordained by the Crown it normally had to be purchased from landowners, often at a price agreed by local juries.[3] Only during a short period in the first two decades of the fourteenth century did they have the benefit of free and unrestricted access to woodland in the Crown manor of Calstock. Timber was a vital material for the construction of buildings, boats, carts, machinery, and tools, for providing supports and shoring for shafts and adits, and for launders and the pumps required to effect drainage. Wood, in both its natural state and as charcoal, was also the principal fuel used for the smelting of ore and refining of silver. Unsurprisingly, expenditure on fuel mirrored the changing levels of annual production recorded in the Exchequer Accounts.

Woodland management

The Bere Ferrers mines called upon some woodland that belonged to the Crown, but primarily it was using that owned by the Abbots of Buckland and Tavistock (see Figure 3.5). Unlike small landowners who may not have had lasting, multi-generational production as a priority, these institutional landowners, by their nature, saw woodland as a long-term financial resource. Although there is no unequivocal documentary evidence for the coppicing of woodland in the Bere Ferrers region this does not mean that it was not employed, and men such as Hugo de Anythel, keeper of the wood at 'Warle' (Warleigh) in 1313–17,[4] were specifically employed to oversee the management of woodland used by the mines. Increasing population levels, the development of industry, and expansion of agriculture in the high middle ages saw the need to safeguard supplies of wood through stricter management regimes and this was to be regulated by statute in the sixteenth century.[5] Coppicing is believed to have been widely practised by 1086, and by the mid-thirteenth century was almost ubiquitous in British woods, although the surviving record is biased towards eastern England.[6] Until recently, however, the archaeological evidence for woodland management was sparse, although paleoenvironmental material from the Rievaulx/Bilsdale area of North Yorkshire now shows that well-managed woodland could sustain significant iron production without resorting to wholesale felling.[7] A number of strategies could be employed which varied from restricted yearly cycles for cropping underwood and small poles, a seven to ten yearly cycle for the production of larger poles and wood for fuel, and over two full cycles to allow the growth of 'standards' (larger trees used for structural purposes). For the post-medieval period Crossley has suggested that charcoal from 7 to 12 year old wood was the fuel of choice for an element of the lead smelting process.[8] Such was the value of woodland, particularly in the long term, it is said that Henry III and his successor Edward I saw large timber-producing trees as capital.[9] The same division between timber-producing trees being regarded as capital while coppiced trees were an asset that provided regular income has been discussed in relation to the woodland of Battle Abbey in East Sussex.[10]

At Bere Ferrers the returns from woodland management were certainly maximised through the consumption of all usable products. Brushwood was used to fuel the smelting boles, while small wood was sold for domestic consumption. Oak trees felled for use in the mines were stripped of their bark that was sold locally to the tanneries. Oak bark, high in tannins, was a key material used in the tanning industry for the preservation of leather, and in 1309–10 a hundred seams of bark sold to the tanneries brought an income of 25s.[11] Significantly, the ash of the bark, once used for tanning, was bought back from the tanneries to be used in the silver refining process, and the purchase of refinery ash, as it is known, is recorded on various occasions between 1294 and 1481, with the majority being transported from Exeter although some was bought at Buckland, 'Yodbira' (Rodborough) and Plympton.[12] Although the mine accounts allow us an insight into the locations of woodland used for the mines and the products that it supplied, we suffer the common problem that there is insufficient evidence – be it archaeological, ecological, or documentary – to reconstruct specific regimes of woodland management or its relationship to silver mining.[13]

The Cistercian monastery at Buckland would have no doubt had a strict regime of management employed at their woods of 'Biccombe', 'Blakstone' (Blaxton) (Figure 5.1), and Buckland before the Crown was given the rights to these for use in the mines. Similarly, the Crown's own woodland in Calstock would probably have already been managed prior to 1301 when it was granted to the mines. It is inconceivable that the Crown would allow its managed woods to be over-exploited, and King Edward I's attitude to the value of woodland suggests that if there was insufficient that could be cropped without long-term disruption or damage, then woodland products would be purchased from elsewhere or the rights to other woodland leased. Neither could the Crown countenance the unreasonable wastage of woodland by the keepers of the mines. After the departure of the Frescobaldi in 1300, the complaints of the Abbot of Buckland that

Figure 5.1:
Maristow (centre) and Blaxton Wood (right) looking across the Tavy estuary from Bere Ferrers. 'Martinstow' (Maristow) was one of the administrative centres of the royal mines and also a smelting centre (see Figure 3.5 for locations).

PHOTOGRAPH: STEPHEN RIPPON

at least 300 acres of woodland had been devastated by the mine, with the allegation that half of the damage had been done during the Frescobaldi's tenure, were upheld by a local jury and the Crown pledged that it would not happen again.[14] The attitude of management under the Frescobaldi would appear to have been much as it was to ore production, achieving a quick return without regard for future production (Chapter 3 above), and ran counter to the general evidence for long-term woodland management.[15]

In terms of mining, unlike the tenure of the Frescobaldi, the Crown had the goals of long-term production which would have required sustainable sources of wood for fuel and timber for building. It has been said that coppicing was the only way to sustain the medieval iron industry,[16] and that is supported by Wheeler's conclusion that wood stocks at Rievaulx and Bilsdale in North Yorkshire 'were sustained by selective and rotational management to maintain equilibrium between natural regeneration and human exploitation'.[17] This is also likely to have been true of the lead/silver mines in Devon, and the Abbots of Buckland and Tavistock, as well as John de Ferrers, none of whom had direct capital gain from the produce of the mines, would have been opposed to the complete clearance of their woodland as once it was diminished there was no future income to be made. In a recent study, Cannell suggested various mechanisms for limiting the impact on woodland of an industry with a high demand for timber and fuel, including refusal to supply, the use of a variety of different woodlands, and the coordination of rotation and output from these different places to spread the effect of increased consumption.[18] Industry rarely caused the complete disappearance of woodland; in fact more medieval woodland survives in areas such as the Weald and Forest of Dean than in those that saw increased agricultural development.[19] The areas of woodland named in association with the mines, that near Denham Bridge, Morwellham, Blaxton and Warleigh for example, are all wooded today, suggesting that they were never fully cleared but have been managed over many centuries.

As was the case with the pottery and tile industries, which were sometimes situated within or close to woodland,[20] the processing of silver-bearing ores had a close spatial relationship with its fuel supply, and so when Buckland Abbey's woodland on the east bank of the river Tavy at Maristowe became exhausted the focus of ore smelting and refining of silver moved to Calstock where the Crown's own woodlands were assigned for use by the mines. On 26 February 1301 the sheriff of Cornwall was ordered

> to permit Thomas de Swaneseye, king's clerk, keeper of the king's mine in that county … to take from the king's wood of Calistok, in that county, as much wood and as often as shall be necessary for the works of the mines, as the king has assigned the wood for the works of the mines.[21]

These woods in turn appear to have become exhausted, and while in 1315 the keeper of the mines was prevented from using the abbot's woods at Buckland,[22] in the early 1320s it was decreed that when the Crown woods at Calstock were depleted, the keeper could call on the abbot's woods at 'Biccombe' and 'Blakstone', and those of William de Ferrers in Bere Ferrers itself, whereafter smelting and refining returned to Maristowe (Figures 3.5 and 5.1).[23] Later in the fourteenth century the mines also called on 60 acres of John de Ferrer's wood at 'Halsere', either side of Denham Bridge in the east of Bere Ferrers parish and 30 acres of the Abbot of Tavistock's woodland at Morwellham,[24] although compared to the 300 acres of the Abbot of Buckland's wood at 'Biccombe' used by 1301,[25] they were too small to influence the location of the smelting centre. These depleted woods were not, however, completely and permanently cleared – indeed, many of the steeper valley sides in this area are still wooded today – and through co-operation and co-ordination these woods were able to meet the demands of the Bere Ferrers silver industry through the late thirteenth and fourteenth centuries as we have seen in the shift from the eastern side of the Tavy (around Maristowe), across to Calstock, and then back the eastern side of Tavy (at Biccombe).

Charcoal production

Archaeological and documentary evidence concerning medieval charcoal production and its links with contemporary industries is scarce: on the Bere peninsula, and elsewhere, it is more common for the physical traces of the industry which the charcoal fuelled to be recognised and investigated, than the sites which produced the charcoal in the first place. While it is possible to identify the sites of charcoal production on the ground, it is rare for them to be closely dated as their form changed little over time and sites were frequently re-used. The documentary record of charcoal burning in medieval England is dominated by that carried out in the royal forests and on monastic estates, and in particular where it was linked with the iron industry. For example, such was the demand in the Forest of Dean that in 1237 Henry III limited the tree species that could be burned to those less valuable such as thorn and hazel,[26] while in 1290 it was decreed that no timber, brushwood, or charcoal was allowed to leave the extensive woodlands of the Weald of Sussex and Kent because of the heavy demands of local iron production.[27] Monastic houses across medieval England were closely involved in the supply of wood and charcoal, notably in West Yorkshire where Fountains, Rievaulx, Kirkstead, Byland, and Salley Abbeys had rights to mine iron ore and use wood for fuel.[28] Closer to Bere Ferrers, Dunkeswell Abbey in east Devon had an interest in iron production in the Blackdown Hills.[29]

Charcoal was produced by the controlled burning of cut wood in an oxygen-limited environment commonly known as a 'clamp'. The clamp normally took one of two forms: a closely stacked woodpile built on a platform and then covered by earth with holes to allow limited airflow, or a pit which was then sealed. While the latter technique was in use during the medieval period across continental Europe, there is little evidence that it was used in England where the platform technique prevailed.[30] Archaeologically, charcoal burning using an above-ground clamp commonly leaves behind a levelled platform, stepped and revetted into the slope of the ground if necessary. A layer of fine charcoal would be expected.

Considering the significance of the silver refining at Bere Ferrers, there is surprisingly little evidence for charcoal production in the immediate area: there is no documentary evidence for the purchase of charcoal from either Tavistock or Buckland Abbeys, while no charcoal-making platforms have been identified in Bere Ferrers parish. There is, however, evidence from slightly further afield, including from within woodland that are documented as having been exploited by the mines (Figure 3.5). Charcoal platforms thought to be medieval or post-medieval in date have been identified in Great North Wood, adjacent to the river Tavy in Buckland Monochorum parish.[31] The location of Great North Wood, only 1 km north-west of Buckland Abbey, correlates with a description of woodland owned by the Abbot of Buckland and leased to the mines. Further charcoal platforms, measuring up to 5 m by 3 m and stepped into a steep west-facing slope, have been recognised above the river Tamar south of the post-medieval mine complex called Devon Great Consols, about 5 km west of Tavistock, that may be the Abbot of Tavistock's woodland at Morwellham, 30 acres of which were leased to the mines.[32] There are two place-names within Bere Ferrers that are possibly linked with charcoal production: Braunder Wood (Old English *brand* = burning[33]) and Colwill Field recorded in the parish Tithe survey (Old English *col* = either coal or charcoal[34]). Colwill Field is located on a sloping valley side, and comparably steep slopes elsewhere in Bere Ferrers remain wooded.

In addition to production from woodland owned or leased by the Crown, charcoal for fuel was also purchased from secondary suppliers for a fixed price. For example, 213 quarters[35] were bought from a place referred to as *Hankelake* in 1343[36] and 526 quarters from a place called *Hacche* in 1344.[37] Judging from the low cost of transporting charcoal from these places they appear to have been local, minimising the joint problems of bulk transport and also of moving friable charcoal long distances without excessive damage. In the same years charcoal totalling 963 quarters was produced from wood already purchased for the use of the Bere Ferrers mines,[38] showing that almost half of fuel charcoal was bought as a finished product. In 1349 20 quarters of charcoal were purchased from 'diverse places',[39] which suggests that it was bought from a variety of small producers rather than a single large-scale operation. The Exchequer accounts make no record of charcoal produced from wood already purchased for the mines in that year, implying that local woodland may have been laid to rest or that its exploitation had been disrupted by plague. The years of 1343, 1344 and 1349 are exceptional as in no others is it accounted that charcoal was bought as a finished product, which raises that possibility that in this decade there was a local shortage, although organisational changes in how the mines were managed mean that between 1349 and 1361 there is then a break in the mine's accounts. The low quantities of charcoal recorded for the year 1349 indicates a reduction in smelting and refining, which ties in with the fact that there was no mining in that year, only reworking of residues. The mine's returns for that year totalled just £37 2s. 0d.,[40] the lowest financial yield since the initial return of £30 during the Crown's first year of operation in 1292.[41]

Charcoal was the primary fuel used for the smelting furnaces, while the bole hearths used wood. Charcoal also fuelled the re-smelting of slag (or blackwork) and the refining of silver, and was sometimes used to fuel the hearths of the smithies which would have supplied tools and structural furnishings to the mines. Iron and steel were both purchased for the mine, mainly through the port at Sutton Harbour, in Plymouth, but also from Bodmin in Cornwall. Moor coal (peat), brought down from Dartmoor was the main fuel for general smithing in the early fourteenth century but, where high quality finished work was required, mineral coal (sea coal) was imported.[42] The properties of mineral coal as a quality fuel for smithing iron tools were well known by that date. Its advantage lay in its physical properties – its ability to provide a sustained and concentrated heat source capable of withstanding the rigours of the smith's hearth. Particular grades of coal in south-west Wales are noted by sixteenth- and seventeenth-century writers as being preferred by smiths for their ability to combine and consolidate in the hearth. Hatcher lists many instances of the preference for coal and also comments on the problem of charcoal being displaced by the blast from the smith's bellows.[43] Charcoal had replaced mineral coal by the 1340s, probably on grounds of cost.[44] Mineral coal, then referred to as 'stone coal', was again being used in 1480–81.[45]

Like other commodities such as tallow (for candles), rope, and timber, sea coal was shipped by barge from 'Sutton'[46] (probably Sutton Harbour in Plymouth), where the Tamar meets the English

Channel. There is no record to suggest that silver was transported to the exchange in London via this busy port; instead it was transported overland (see Chapter 4). The readiness to employ alternative sources of timber and fuel indicates the value given to sustainable woodland exploitation in the local area. In a period where there were higher levels of demand put upon woodland it may have been commonplace to substitute other fuels.[47]

The wage rolls provide the names of five charcoal burners employed in 1306 (Luke de Crenbere, Walter Cuperon, Reginald Jolyfs, Roger atte More, and Thomas Wytling),[48] though it is not known which woods they targeted for production. Although these are the only recorded charcoal burners in the Bere Ferrers wage rolls, the fact that charcoal for fuel was also bought as a finished product at a higher cost per quarter than that produced from wood already purchased for use by the mines implies that there were others engaged locally in this industry but not employed directly by the Crown. Some charcoal burners, like Luke de Crenbere and his contemporaries, were directly employed though it is believed that the majority were autonomous and combined it with other activities such as agriculture or metalworking.[49] Entrepreneurs were charged for licences to make charcoal in the royal forests,[50] and it is probable that the same occurred in the woods of ecclesiastical and secular landowners. The growth of the silver and lead industry and the associated increased demand for charcoal would have acted as a draw for skilled charcoal burners in the same way that iron production did elsewhere in Britain.

The requirement for water power

One of the uses of timber within the Bere Ferrers mines was in the construction of drainage adits, which leads us onto another facet of this landscape: the management of water. In one sense water was a problem, flooding the mines and leading to the need for drainage, while in another sense it was an asset in providing a source of power. The management of water within the mining landscape of Bere Ferrers gave rise to one of the most remarkable surviving examples of medieval technology: the 16 km long Lumburn Leat.

Drainage

Drainage of water from the workings was a necessity from the start of mining operations at Bere Ferrers. In the thirteenth and fourteenth centuries drainage was achieved by manual haulage. Though the Crown was responsible for the capital costs of providing materials, until the end of the first decade of the fourteenth century the miners themselves had to provide the labour. After this date the Exchequer Accounts record significant wage bills for water hauliers, rising from fourteen men at a cost of c.£6 in 1308[51] to twenty seven men at a cost of £20 9s. 6d. in 1313.[52] Water was raised using leather buckets on ropes, though windlasses and wooden 'kibbles' were brought into use at a later date. Water 'winding', implying the use of a windlass, is not referred to until 1480,[53] although windlasses/winches were in use during trial copper extraction at Dyserth in Flintshire (north-east Wales) in 1303[54] and so it is possible that they were in use at Bere Ferrers in the fourteenth century.

As has already been discussed in Chapter 4, the need to work the increasingly deep silver-bearing deposits gave rise to a constant battle with water entering the workings. The Crown was quick to overcome this and minimise the needs of manual water haulage by constructing drainage galleries or adits to allow water to freely drain to the surface at a lower level. By the beginning of the fourteenth century they had already allowed for working through the wet winter months and reduced the expenses of manually hauling with buckets.[55] The effective use of adits was limited by the topography and as the working increased in depth the driveage required became significantly greater. Drainage using a combination of adits and manual haulage through shafts and galleries was the mainstay of fourteenth-century water extraction, although the high wages after 1348 meant that by the fifteenth century the cost of driving longer adits became prohibitive. The requirement for manual water haulage therefore increased substantially such that, by the middle of the fifteenth century, the cost had allegedly reached £120 over an eighteen week period (although this should be treated with caution as the original indenture has not survived and we are relying on a transcript dated 1608).[56]

Fifteenth-century innovation: the introduction of water-powered suction-lift pumps
The problem of increasing drainage costs did not result in an end to mining on the Bere Ferrers peninsula as may have been expected, because other influences demanded that they were kept operational. During the mid-fifteenth century northern Europe suffered a crisis in bullion supply and the demand for silver encouraged further working

Fuelling the industry

A—Sump. B—Pipes. C—Flooring. D—Trunk. E—Perforations of trunk. F—Valve. G—Spout. H—Piston-rod. I—Hand-bar of piston. K—Shoe. L—Disc with round openings. M—Disc with oval openings. N—Cover. O—This man is boring logs and making them into pipes. P—Borer with auger. Q—Wider borer.

A—Shaft. B—Bottom pump. C—First tank. D—Second pump. E—Second tank. F—Third pump. G—Trough. H—The iron set in the axle. I—First pump rod. K—Second pump rod. L—Third pump rod. M—First piston rod. N—Second piston rod. O—Third piston rod. P—Little axles. Q—"Claws."

Figure 5.2: Two woodcuts, taken from the work of Georgius Agricola in 1556, illustrating the construction and assembly of suction-lift pumps much as it would have worked at Bere Ferrers a hundred years earlier. At Bere Ferrers the pumps would have been powered by water wheel, rather than hand operated, in a manner similar to that in the right hand illustration.

IMAGES SUPPLIED BY STEPHEN HENLEY

of the deep-seated deposits on the Bere peninsula, and drainage at the northern end of the mines, in the area around Lockridge Hill, was addressed by the introduction of mechanised pumping. The introduction of water-powered suction-lift pumps (Figure 5.2) was a landmark time in terms of the adoption of innovative technology and hydrological engineering in Devon. The available documents are silent as to the precise location of the pumps, but the focus of mining in the mid-fifteenth century was on the northern part of the mines in the

Figure 5.3: Route of the Lumburn leat shown against a background of the Ordnance Survey First Edition Six Inch map.

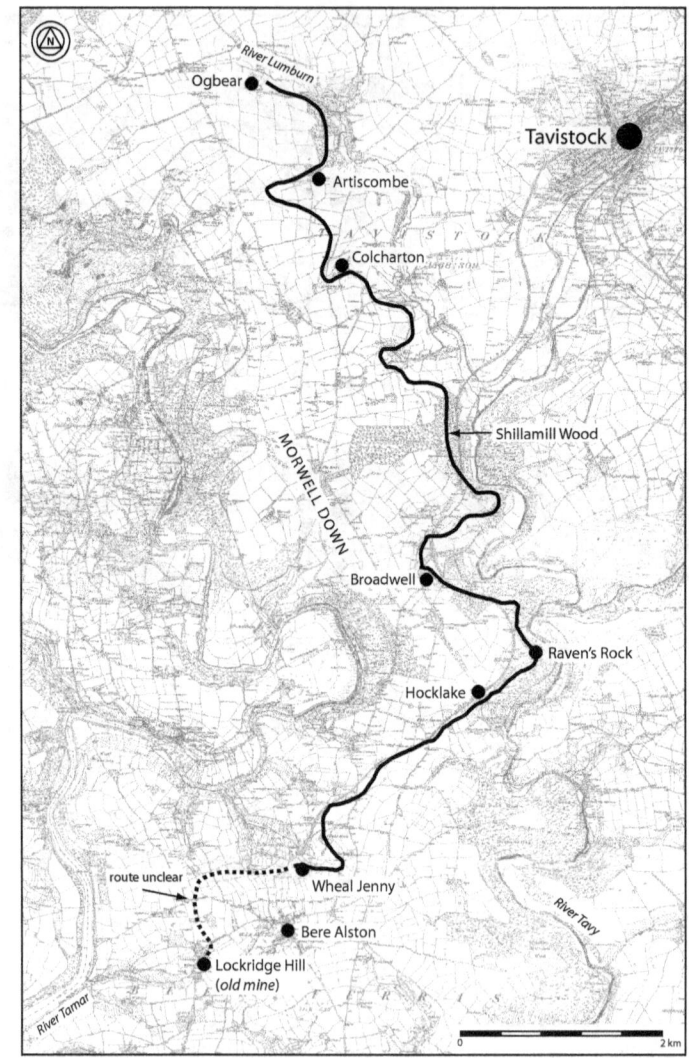

area of Lockridge Hill, with a group of shafts there being reserved to the Crown when others were leased out to entrepreneurial interests (Figure 3.7). The pumps, driven by a water wheel and probably using cranks for vertical power transfer, required a constant and regular supply of water to power them. The topography and hydrology of the area meant that there was no locally available water supply which flowed from a higher level than the pumps, and so the only answer was to divert and carry water over a distance of sixteen kilometres from Ogbear, west of Tavistock, via a purpose-built leat, the scale of which appears to be unique in medieval England (Figure 5.3).

The suction-lift pump

The woodcut in Figure 5.2 (left), taken from the work of Georgius Agricola in 1556, illustrates the construction and assembly of a suction-lift pump much as it would have worked at Bere Ferrers a hundred years earlier, except that it would have been powered by water wheel rather than hand-operated.[57] The barrels of the pumps, at least nine of them with a total length of 20 fathoms (36 metres), were bored from trees purchased at Plympton Wood, about 15 kilometres to the south-east, and probably brought to the mine by river barge. There they were assembled as an ascending series in the shaft in much the same manner as they are in another of Agricola's woodcuts (Figure 5.2 right), capable of lifting water up to a point were it could freely drain to surface, probably along an adit. The whole assembly, water wheel and pumps, was referred to as the *ordenance*: 'to Richard Petersfield for looking after the ordenance there both by day and night'.[58]

Operated by the water wheel, the piston and piston rod reciprocated within the barrel of the pump. Upward action of the piston in the pump drew in water from the sump through a flap valve while the return action downwards allowed the water to move, through another flap valve, to the upper side of the piston, and a leather hide was purchased for the *sowkers* (i.e. suckers) at the *ordenance* in 1480–81.[59] Further upward action closed the valve in the piston, drawing in more water while lifting that above the piston toward the top of the pump barrel. Continued reciprocating action resulted in the water being lifted to flow out of the top of the barrel either to the next pump or to flow out to surface. The height to which such a pump could lift water was only restricted by the mechanical strength of its component parts and the power of the water wheel driving it.

The suction-lift pump was truly innovative technology. Developed in Italy in the first half of the fifteenth century, it was in use in the central European silver mines by the 1450s, although the route by which it reached Bere Ferrers is unknown.[60] Sir John Fogge was the lessee of the mines from 1471 onwards and, before that, treasurer to Edward IV's household. He had the opportunity for contacts with central European mining interests while in the king's service in England and during the latter's period of exile in Burgundy.[61] Soon after he was granted a lease of the mines it was stated that 'it was agreed that the said John should bring water to the mines at his own expense'.[62] So within twenty years of their adoption in central Europe Fogge appears to have had it in mind to introduce the pumps to Bere

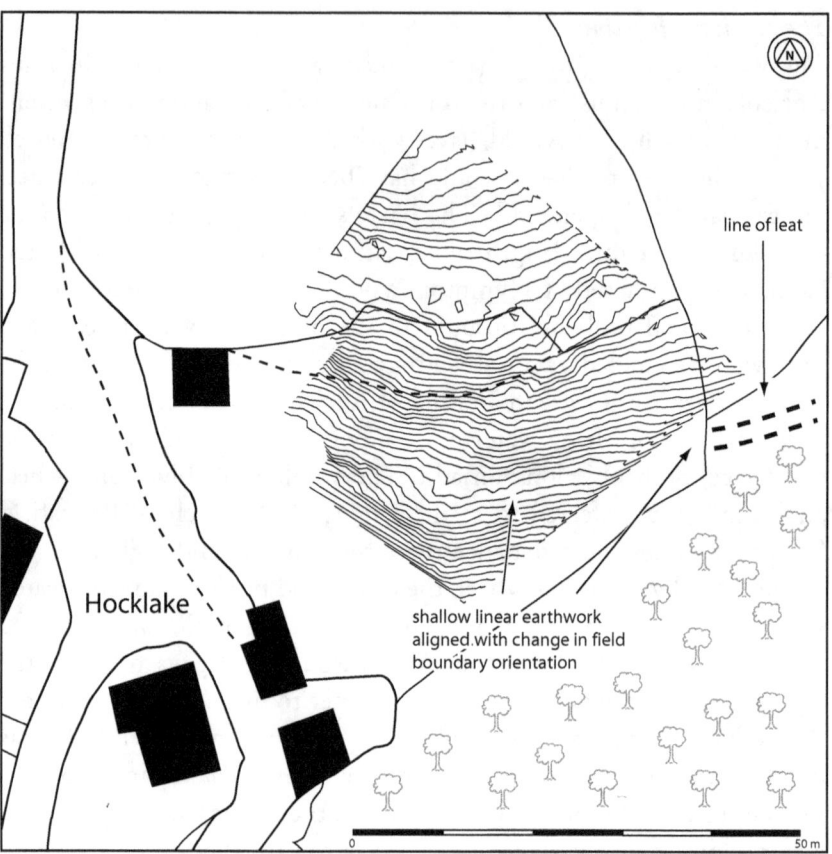

Figure 5.4: The line of the Lumburn Leat at Hocklake Farm, produced through a differential global positioning system survey where elevation readings were taken over an area of 45 m by 40 m at 1 m intervals producing a high-resolution contour plan of the site, mapped at a 0.01 m interval. Though it was not perceptible in the field, a distinct break of slope was revealed, the elevation of which matches surviving sections of the leat to the north and south.

Ferrers and was constructing the leat that would provide them with their motive power.

Before the start of the Bere Ferrers Project, little was known about this leat. Several short stretches of it were identified by Booker in the 1960s[63] but he was unable to determine their purpose, and while Jenkin[64] noted the documentary evidence for water-powered pumping in the mines, he did not make the connection between the two. Research in the 1990s by one of the present authors,[65] however, suggested a link between the documented pumps at Lockridge Hill and these physical remains, and fieldwork undertaken by the Bere Ferrers Project was designed to test this hypothesis. A differential global positioning system (dGPS) was used to take accurate height readings at regular intervals along the full length of its course, confirming that surviving stretches of leat form a single continuous feature, gradually declining in elevation from source to end. Over a substantial part of its course, between Artiscombe and Wheal Jenny, it has a fall of little over 18 metres, an overall gradient of about 1:800. Furthermore, the use of dGPS allowed the probable line of the leat to be traced where it no longer survives as an earthwork. One example of this work was a gently sloping paddock at Hocklake Farm where elevation readings were taken over an area of 45 m by 40 m at 1 m intervals producing a high-resolution contour plan of the site, mapped at a 0.01 m interval (Figure 5.4). Though it was not perceptible in the field, a distinct break of slope was revealed, and not only does the elevation of this feature match surviving sections of the leat to the north and south, but it also corresponds to a slight change in the line of the adjacent field boundary. It several places where the leat is not recognisable as an earthwork feature running through woodland or across an open field, its line is marked by a field boundary.

The source of the leat's water was the head of a tributary of the river Lumburn (at SX443751). Ephemeral earthworks at this location have the appearance of water-holding tanks, used to manage the supply of water along the leat. The channel survives in various states of preservation as an earthwork feature, with the most complete sections being those which run through woodland on the sloping and wooded valley sides (Figure 5.5 and 5.6). In such places it is on average about a metre wide and half a metre deep, though a certain amount of infilling

Figure 5.5: The Lumburn Leat at Broadwell, downslope from the Crowndale to Orestocks road.
PHOTOGRAPH: PETER CLAUGHTON

Figure 5.6: The Lumburn Leat in Shillamill Wood.

PHOTOGRAPH: CHRIS SMART

BELOW LEFT

Figure 5.7: A deep cutting for the Lumburn Leat at Raven's Rock.

PHOTOGRAPH: PETER CLAUGHTON

BELOW RIGHT

Figure 5.8: Tunnel on the Lumburn Leat at Raven's Rock.

COPYRIGHT CAROLINE VULLIAMY

Fuelling the industry

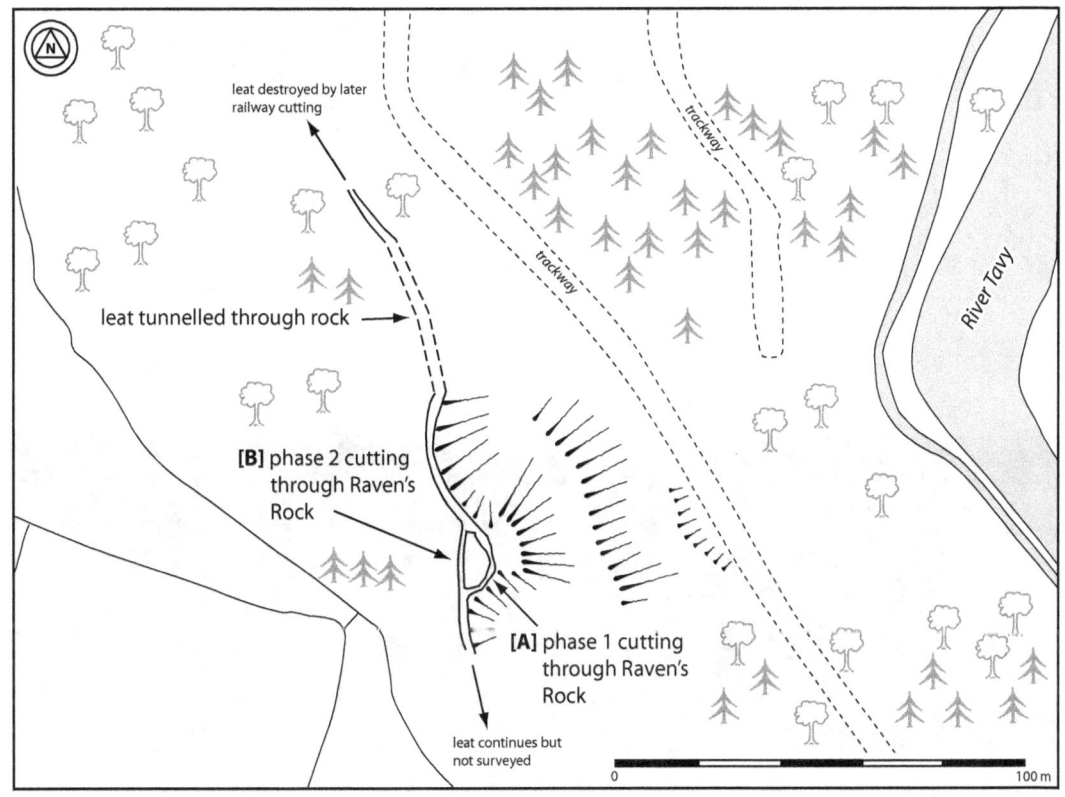

Figure 5.9: Course of the Lumburn Leat at Raven's Rock.

masks its true profile. It is also possible, however, that in places it has been modified through later re-use as an internal woodland boundary. In places the leat's engineers came up against steep rocky outcrops high above the river Tavy, and here it was necessary to tunnel through the hillside, and these cuttings survive intact at Raven's Rock and in Shillamill Wood (Figure 5.7 and 5.8). Where the leat encountered the two outcrops at Raven's Rock it is possible to identify modifications to its course that were either made during its initial construction or over its lifetime (Figure 5.9 and 5.10). At the first outcrop there appear to have been initial attempts to cut a ledge around the spur, perhaps to carry a wooden launder. This apparently never worked, or failed at some later point, because a tunnel was then hand-cut through the outcrop on a more direct line. Two opposing slots at the entrance to the tunnel, still showing the pick marks of those that dug it, are thought to have been runnels for sluice boards put in place to control flow. Approximately 100 m further south the leat's path was blocked by another outcrop. Here, an earlier cutting was made down through the

Figure 5.10:
The two cuttings for the Lumburn Leat south of Raven's Rock (for plan see Figure 5.9).
PHOTOGRAPH: PETER CLAUGHTON

rock on the outer edge, but again a launder, which may have brought it back to the hillside, probably failed, so a second cutting was made through a greater depth of rock on a more direct course.

The final stretch of leat, as it is routed to the north and west of Bere Alston towards Lockridge Hill, has proved more difficult to identify. The construction of a railway line in the late nineteenth century, and the improvement of agricultural land, have removed virtually all traces of its course. The height of the leat at its last surviving stretch, to the rear of Wheal Jenny Bungalow, and at its destination, Lockridge Hill, are known, giving upper and lower limits of its elevation between the two. This allows the corridor through which the leat must have past to be identified, and this reveals that it must have crossed through an open-field to the north of Bere Alston (discussed in Chapter 6).

The construction of this leat was a major undertaking, but the returns from this investment were not as substantial as may have been hoped as it became clear that the deeper productive areas were further south, under the river Tamar. Production of silver during 1480–81 reached just 1,883 ounces,[66] one of the poorer yields and quantity of silver per weight of mined ore recorded in the surviving mine returns. By the early sixteenth century the accessible silver-bearing deposits were essentially worked-out and mining ceased. The nature of the mineralisation, and the ability of the medieval miner to work at considerable depth, meant that it was not until the early nineteenth century and the availability of efficient steam-powered pumps that it was possible to get under the 'ancient bottoms' – the late medieval limits – up to 50 fathoms below surface in places, and effectively work the southern ground.[67]

Smelting and refining

It was not until the late fifteenth century that water power was harnessed to aid mechanised drainage, but it had long been used in the processing of ore and refining of silver. As was discussed above, processing shifted between Maristow in Devon and Calstock in Cornwall in accordance with the availability of wood for fuel. One or more sites at Maristow, in the Tavy estuary were used for the smelting and refining of silver-bearing ores from the Bere Ferrers mines. The site was originally chosen as the site of the water-powered refining mill in 1292, activity which was transferred to Calstock in about 1302. In addition to the ore from the Bere Ferrers mines, it is recorded that fertile lead (lead-bearing silver) was being transported to Maristow from the mines at Combe Martin for refining in the 1290s. By the 1320s refining was again being carried out there along with smelting by furnace.[68]

The bellows that provided the blast for the refining mill of the 1290s were water-powered, as were those of the later, fifteenth-century 'fynyngmyll'. The smelting furnaces were probably powered in a similar way. It is unlikely that the river Tavy provided the source of this water power as its tidal reaches continued some way upstream of Maristow (the Lopwell Dam being a recent feature). The refining mill was certainly some distance from Maristow itself as, when the mill was dismantled in 1303, a total of 8d. was paid for moving the bellows to the river for shipping to Calstock.[69] This sum would normally constitute eight man-days in labourers' wages, although this heavier work may have attracted a higher rate.[70] Accordingly, the most likely source of power would have been the Milton Brook, which entered the Tavy at Lopwell. At present it has not been possible to prove the precise location of the sites at Calstock or Maristow, though suggestions were made in Chapter 4.

Discussion

The Bere Ferrers silver mines led to an unprecedented demand for wood for fuel as well as timber for construction. For the early fourteenth century the detailed mine accounts detail the purchase of wood, timber, charcoal, sea coal, and peat, which reveal how the silver industry was fuelled. With the exception of John de Ferrers' woodland at 'Halsere' adjacent to the river Tavy near to Denham Bridge, there is no archaeological or documentary evidence for woodland within Bere Ferrers parish being used for the mines. It was the Crown's own woodland at Calstock, the Abbot of Tavistock's woods at Morwellham, and the Abbot of Buckland's wood and timber resources on the other side of the river Tavy that supplied the medieval silver industry. There is no documentary evidence for the management of any woodland under a system of coppice rotation, although contemporary accounts of other medieval woodland, the on-going demands for wood and timber, and the evidence that wood was sourced from different locations as it became exhausted elsewhere, provides a strong indication that it was practised in this area. Monastic landowners, as well as secular lords such as the de Ferrers family, would have benefited from the sustained income from managed woodland and the capital assets of standards left for timber. It was the woodland resources that provided the source of heat to smelt the ore and refine the silver, but water power was harnessed to drive the bellows which provided the blast for the furnaces. Water would have been drawn from streams entering the Tavy and routed via leats to the water wheels at the furnaces and the refining mill. Until the late fifteenth century, drainage of the mines was done by manual haulage using buckets and rope, but then mechanised pumping using innovative suction-lift technology was introduced. This demanded the construction of a remarkable leat which ran for over sixteen kilometres, the scale of which is unparalleled in medieval England.

Chapter 6

The mining community and its impact on the wider landscape

During their heyday, the silver mines at Bere Ferrers gave employment to hundreds of workers. Other than the miners involved in the actual extraction of ores, they included water hauliers, smiths, carpenters, sawyers, chandlers, smelters, refiners, wood cutters, charcoal-burners and carters.[1] It is recorded that at least 400 men were drafted in from other mining districts across Britain, including the lead mines of the Peak District and the tin-workings of Dartmoor and Cornwall. Such an influx of people from outside the local area would have had led to a significant demand for food, accommodation and other amenities, and this chapter examines the impact that this mining community had on the wider landscape.

Characterising the historic landscape

Unfortunately, we do not have a medieval map of the Bere Ferrers area, and so we must rely upon later sources. A key concept here is the 'historic landscape', which is a term used for the present pattern of fields, roads and settlements in use today, and is designed to show that these features exhibit great historical interest. In many parts of the country, such as Devon, for example, research has shown that many features within today's countryside and towns date back to the medieval period, and by studying the historic landscape we can start to reconstruct what it looked like in earlier times.[2] Today's historic landscape – as shown on maps and aerial photographs – can be analysed through a technique that is known as 'Historic Landscape Characterisation' (HLC). Various forms of HLC have been developed in England, Wales, and Scotland to study the historic landscape, although these tend to focus on informing planners and countryside managers

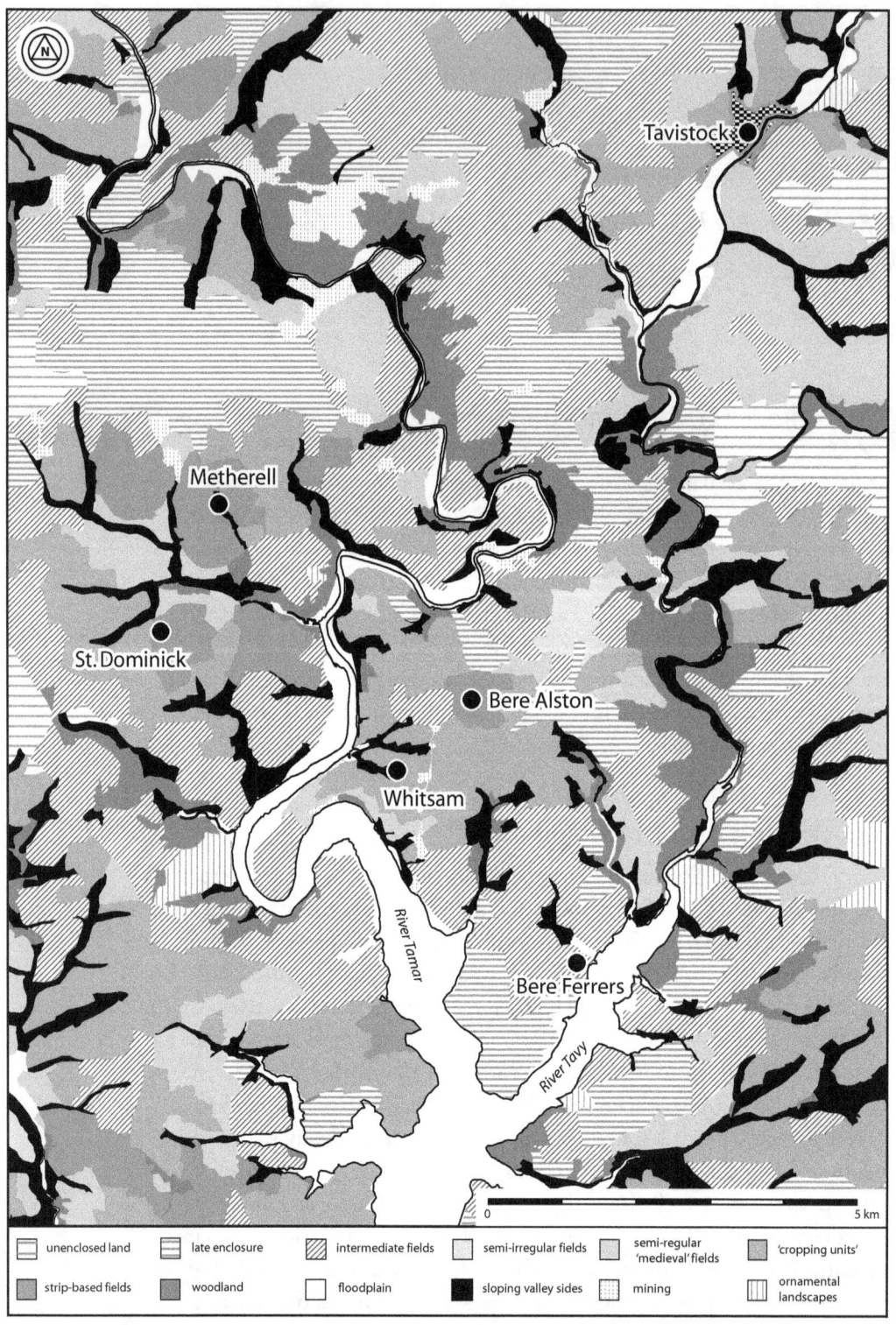

how old our countryside is (its 'time-depth'), and the major processes that have led to the landscape achieving its current appearance. This work has usually been carried out on a large scale – usually covering entire counties – rather than in-depth studies of particular locations.[3] Criticisms of 'Historic Landscape Characterisation' have tended to over-shadow its research potential, which has led to Rippon's use of the term 'historic landscape analysis' to distinguish this wide range of interdisciplinary research-led applications from HLC, the narrower planning and management oriented work that English Heritage, for example, have sponsored across England. In historic landscape analysis, the attributing of different field systems to pre-determined types – that is, essentially what HLC does – is taken further by examining other facets of the landscape such as settlements patterns and the nature of landownership and occupancy.

In this study, an analysis of the historic landscape was required because the current settlement patterns and field systems are of a type that elsewhere in Devon are known to have come into existence by the eleventh century,[4] and so may already have been present when the Bere Ferrers mines were first opened, and have also outlasted them. In order to unravel the development of this historic landscape, the start point was modern Ordnance Survey mapping, available in digital form through Edina Digimap, and the nineteenth-century Six Inch to the Mile maps[5] scanned, geo-referenced (i.e. converted to the same scale as the modern Ordnance Survey mapping), and placed in a computer-based GIS (Geographical Information System). These Six Inch maps were used in a 'historic landscape characterisation' in which the land-use and morphology of the nineteenth-century countryside was assessed, and each parcel of land attributed to one of a series of different 'historic landscape types', such as woodland, floodplain, blocks of long narrow curving fields derived from the enclosure by agreement of former open field, and rectilinear fields derived from the recent enclosure of open pasture (Figure 6.1 and Table 6.1). In the last example, the historical process that led to the creation of this landscape type is known through analogy with other landscapes, and in the case of Bere Ferrers a comparison of the nineteenth-century sources with an earlier map of the manor of Bere Ferrers, drawn up in 1737 for Lord Hobart, reveals the process that led to the creation of this distinctive historic landscape type: field systems characterised by relatively large, straight-sided, rectangular fields. In the nineteenth century, areas of such field boundary patterns were to be found on

OPPOSITE

Figure 6.1: A characterisation of the historic landscape in Bere Ferrers and the adjacent parishes. Based on the Ordnance Survey First Edition Six Inch map. The areas of medieval settlement and field systems concentrate on the gentler, lower slopes, with woodland cloaking the steeper valley sides (e.g. Figure 5.1). the intervening areas of higher ground were generally enclosed later (e.g. see Figures 6.3 and 6.4).

Table 6.1 Landscape character types in the Bere Ferrers region based on the Ordnance Survey First edition Six Inch maps

Character type and description	Visual example
Unenclosed land This character type is typified by rough moorland vegetation and no internal land divisions. Although roads may run along the boundary between this and other character types, no metalled roads run across 'unenclosed land'. It is common for footpaths and tracks to cross these areas. The remains of quarrying and mining are often located in unenclosed areas. On the south-west fringes of Dartmoor there are areas of unenclosed ground in the north of Buckland Monochorum parish and the south of Whitchurch parish. There is also unenclosed land on the slopes of Kit Hill, in the far west of the study area in Cornwall.	
Late enclosure Fields of this type have dead-straight sides and geometric shapes. There is variation in field size within this type, such at St Ann's Chapel in Calstock parish where there is a large area of particularly small late enclosure fields. There are few settlements within 'late enclosure' type landscape, and those that do occur are on its edge. 'Late enclosure' in Calstock parish is an exception to this as there are many habitations dispersed across this type, probably linked to the extensive post-medieval mining industries in that area. Unsurprisingly, the roads which cross these areas are straight, but are also wider than many of the roads winding through other character types. There are extensive mine and quarry remains within 'late enclosure' landscapes. Areas of 'late enclosure' are found across the whole study region most commonly on high ground, particularly on the ridge known as Morwell Down between the top of Bere Ferrers parish and Gulworthy, and also across much of the northern half of Calstock parish in Cornwall. There are also smaller areas interspersed with other character types, but most often adjacent to 'intermediate' enclosures on hill slopes.	Figure 6.3: Tuckermarsh
Woodland There is extensive woodland on the steep slopes of the Tamar and Tavy valleys and their tributaries. This woodland is predominantly deciduous though it is apparent that some broad-leafed woods have had coniferous species planted within them since the late nineteenth century (in Hole Wood, Bere Ferrers for example). It is likely that larger areas characterised as 'sloping valley sides' (see below) were once also wooded. There is no woodland found outside of valley side locations within the study area. Like 'sloping valley sides' the upper boundaries of 'woodland' often respect a particular contour and mark the limit of enclosure for fields of other character types. There are no settlements within 'woodland' though elsewhere there are many places with names associated with woodland clearance suggesting that it was once more extensive (e.g. Leigh and Rumleigh in Bere Ferrers parish). There are many places where the boundaries of woods have been 'cut into' during the expansion of agriculture. In places this was through piecemeal assarts such as at Birch Wood/Middle Tor in Buckland Monochorum parish. Elsewhere the clearance of woodland appears to have been more widespread resulting in	Figure 6.3: Buttspill Wood; Figure 6.14: Comfort Wood; Figure 6.15: Hangingcliff Wood

large portions being lost to more regular 'intermediate' enclosures as at Hatch Wood to the south-west of Gulworthy and probably at Great North Wood to the west of Buckland Abbey. There are frequent mine remains within this character type, and in some cases woodland has been cleared to make way for vast mining complexes, such as at Devon Great Consols in Gulworthy parish and Gunnislake Clitters Mine in Calstock parish.

Floodplain The tidal reaches of the rivers Tamar and Tavy have narrow banks of intertidal floodplain along their shorelines. Further upstream there are flat areas adjacent to these rivers and their tributaries which are similar in character though are freshwater environments. Where enclosed, these areas of pasture and meadow have boundaries which mirror the course of the river, and also have straight subdivisions. In the southern half of the study area floodplains often form a barrier between the river and enclosures of other character types. In this area the valleys are less steep and can be cultivated or grazed down to river level. In the upper reaches the topography is more dramatic and flat areas of floodplain are often overlooked by extensive areas of 'sloping valley sides' or 'woodland'.

Figure 6.3: Calstock

Sloping valley sides The characteristic topography of this region is a series of spurs and ridges dissected by a network of valleys carrying rivers and streams which flow into the rivers Tamar and Tavy. These tend to be steeper and more pronounced towards in the northern half of the study area. This character type consists of the often steeply sloping ground that dips from adjacent character types to the valley bottoms. In most cases the ground is too steep to allow cultivation, though on shallower slopes it has been enclosed and is used as pasture. These enclosures subdividing this character type may be remnants of woodland parcels as woodland still covers many of the steeper slopes. There are some settlements within this character type – mainly farms situated at the heads of valleys though some sit on the junction between the plateau and the valley side.

Figure 6.14: Newhouses
Figure 6.15: Gullytown

Intermediate fields Large areas of rolling upland plateau and gently sloping ground contain what have been classed as 'intermediate' enclosures. These fields are generally large and of regular rectangular, sub-rectangular, and polygonal shape. The boundaries of these enclosures tend to be fairly straight though they do not have the precise geometry associated with 'late enclosure'. This character type is often found adjacent to areas of late enclosure, woodland and sloping valley side, reflecting its plateau and gentle slope locations. Elsewhere it abuts the smaller, more curving enclosures of 'cropping units' (see below). There are possibly subdivisions within this type – for example larger enclosures found in the very north of the study area in Lamerton and Tavistock parishes, both of which were the focus of a medieval manorial centre, one of which, Hurdwick, was the Abbot of Tavistock's sheep rearing grange. There are few settlements within this landscape type and those that occur sit on the boundary between it and an adjacent type, or bear names such as Newhouse (Bere Ferrers parish). It is possible that an area to the west of Millhill of this character was formerly an area of common field cultivation as documents relating to Tavistock Abbey recall the enclosure of strips and furlongs between Newton and Ogbear.

Figure 6.14: Fursdon

Strip-based fields These fields are the enclosed remnants of common field agriculture of medieval date. These enclosures are typified by a long, narrow curving morphology. They occur in congruous blocks although some may be wider than others – a result of agglomeration of open field strips prior to enclosure, reflected in diagnostic characteristics such as dog-legs within boundaries. Although it is possible that fields curve to follow the natural topography, many of these strip-based fields are not oriented to benefit from such an arrangement. The settlements associated with these systems tend to be small nucleations or compact hamlets. Fine examples can be seen surrounding Bere Alston, Metherell, St Dominick and Boketherick. It is evident that these systems were once more extensive but have been subsequently enclosed and rationalised, such as near Ogbear where there is documentary evidence for common field agriculture but little trace today. Smaller blocks of this type can be seen at Cargreen and Carkeel and at Latchley, each on the Cornish bank of the Tamar.
Figure 6.3: Bere Alston
Figure 6.14: Metherell

Cropping units There are large areas of gently sloping ground, particularly at lower elevations, characterised by fields that are longer than they are wide, often with a gently curving profile. These fields often run out laterally from roads which provide a sinuous axis for the surrounding fields. Roads often run between adjacent groups of such enclosures following the curving long axis or the straight short axis of the fields. 'Cropping units' are found in contiguous blocks suggesting that they were created in a single period. Dead-straight subdivisions within these enclosures show adaptation at a later date. In places this effectively creates square fields. The settlement pattern associated with this type is predominantly dispersed farmsteads though some of these may well be the shrunken remains of what once were hamlets. Examples of this type can be seen around Whitsam in Bere Ferrers parish, inland of Cargreen in Landulph parish, around Broadley in Tamerton Foliott parish, and to the north of Buckland Monochorum. This type is very infrequent on the higher ground in the northern half of the study area although one block may relate to the former strip fields documented near Ogbear. There is also evidence for this type around Tavistock.
Figure 6.3: Ashen
Figure 6.14: Trehill
Figure 6.15: Whitsam

Semi-regular 'medieval' fields Enclosures of this type occur in blocks with boundaries that do not always follow the same axis. They tend to have near-square corners and vary in shape from rectangular to square, with others less regular in shape. On occasions boundaries may be curving or have rounded ends. There are few roads within this type which tends to be located on lower slopes. Most often, 'semi-regular medieval fields' are found between 'cropping units' and 'intermediate' or 'late enclosure', and above 'sloping valley sides'. Relative to 'strip-based fields' and 'cropping units', there are few settlements within this type, suggesting that it was perhaps a phase of later medieval enclosure that occurred after the settlement pattern had been established. Examples of this character type are located across the study area, with extensive tracts to the south of Tavistock, between Metherell and St Dominick, and between Landulph and Botus Fleming.
Figure 6.14: Bury

> **Semi-irregular fields** 'Semi-irregular fields' often have roughly geometric shapes with some angular corners but no evidence for an underlying planned layout. The largest extent of 'semi-irregular fields' is situated to the north-east of Bere Alston on a moderate slope between 'semi-regular medieval fields' and 'intermediate fields'. It seems likely that this type represents a phase of enclosure less planned than the intermediate type.
>
> Figure 6.3: north-east of Bere Alston
>
> **Mining** Mines named on Ordnance Survey First Edition Maps have been assigned to this type. In addition to the less-obvious medieval mining industries, there were extensive post-medieval mines across this area. Mining radically altered the character of the landscape and left workings and spoil heaps to be seen on the ground.
>
> Figure 6.3: Lockridge Mine
> Figure 6.15: Furzehill
>
> **Ornamental landscapes** There are only a small number of ornamental landscapes – those that have been adapted for aesthetic reasons – within the study area. They are associated with a number of grand houses such as Buckland Abbey, Cotehele and Pentillie Castle.

some of the higher ground to the north and east of Bere Alston which in 1737 had been unenclosed common land, such as 'Beeralston Down': clearly, this distinctive type of landscape was created through post-medieval enclosure of common open pasture (Figures 6.2–6.4). Other illustrations of this historic landscape characterisation – overlain on the Ordnance Survey First Edition Six Inch maps from which it was derived – can be seen in Figures 6.14 and 6.15.

A key feature of an historic landscape characterisation is that all areas of land are attributed to their dominant historic landscape character types, though for our purposes some – such as ornamental landscapes and urban areas – need not concern us further. By the nineteenth century woodland in the Tamar and Tavy valleys was of limited extent and mostly restricted to the steeper valley sides, though in the not too distant past it may well have also covered other equally steep slopes. Apart from limited areas that in the nineteenth century were still unenclosed – mostly restricted to the higher ground – the remaining areas were enclosed by a series of field systems with very different morphologies. Some are clearly medieval in origin, notably the strip-based fields and 'cropping units', the latter being a distinctive type of field system found in the South West and related to arable farming.[6] Semi-regular fields are also probably of medieval origin, being derived from the piecemeal assarting of woodland and open pasture. Moving to more recent times, quite large areas have distinctive rectilinear layouts

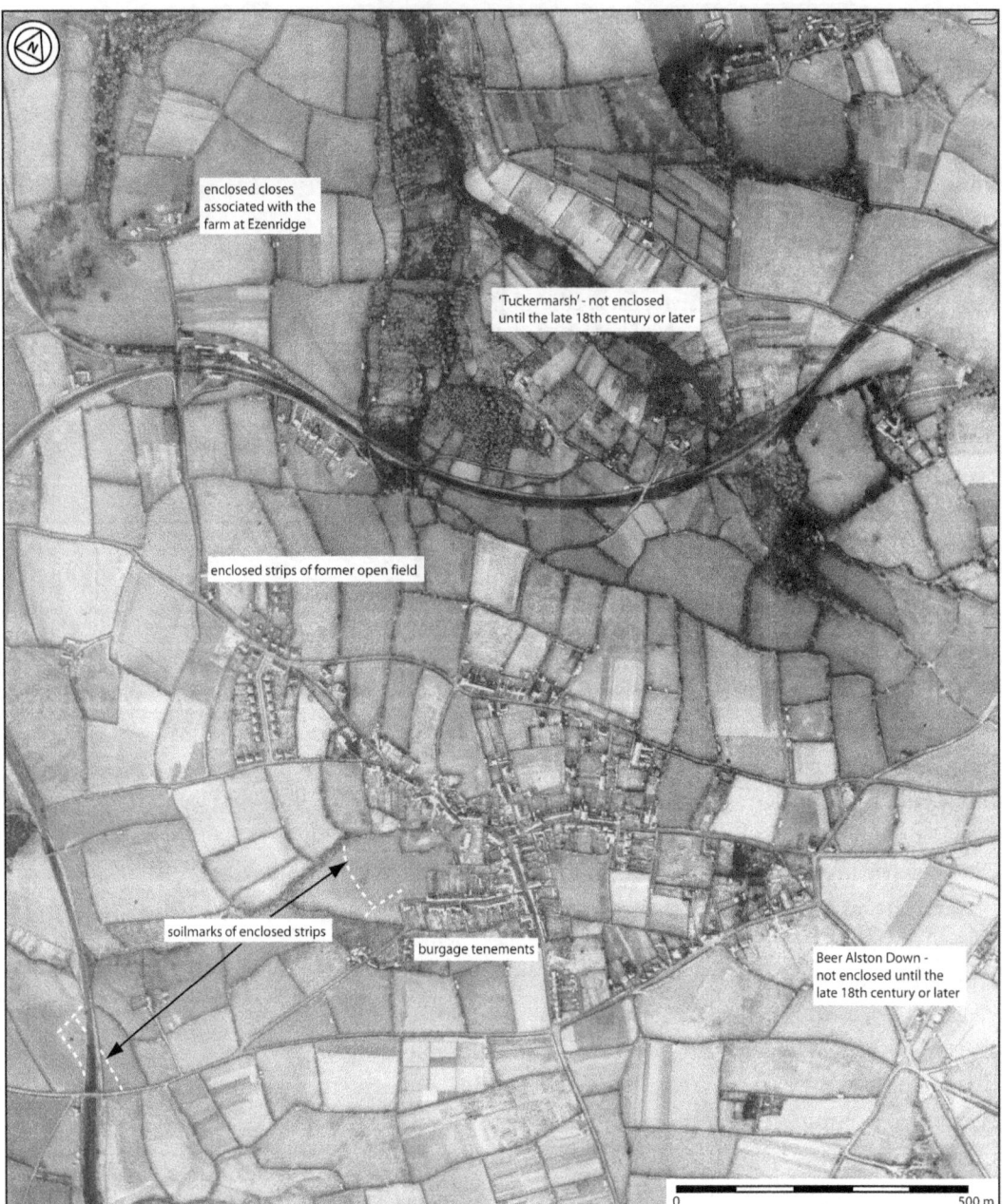

Figure 6.2: Vertical aerial photograph of Bere Alston taken in 1946 showing the planned block of medieval burgage tenements (along what in 1737 was called Pepper Street: see Figure 6.4) to the south what was probably an earlier hamlet at Frog Street, surrounded by a former open field whose strips have been fossilized by post-enclosure field boundaries (CPE/UK/1890: 10DEC '46 F20" // MULTI (4) 54 SQDN. Frame 2341).

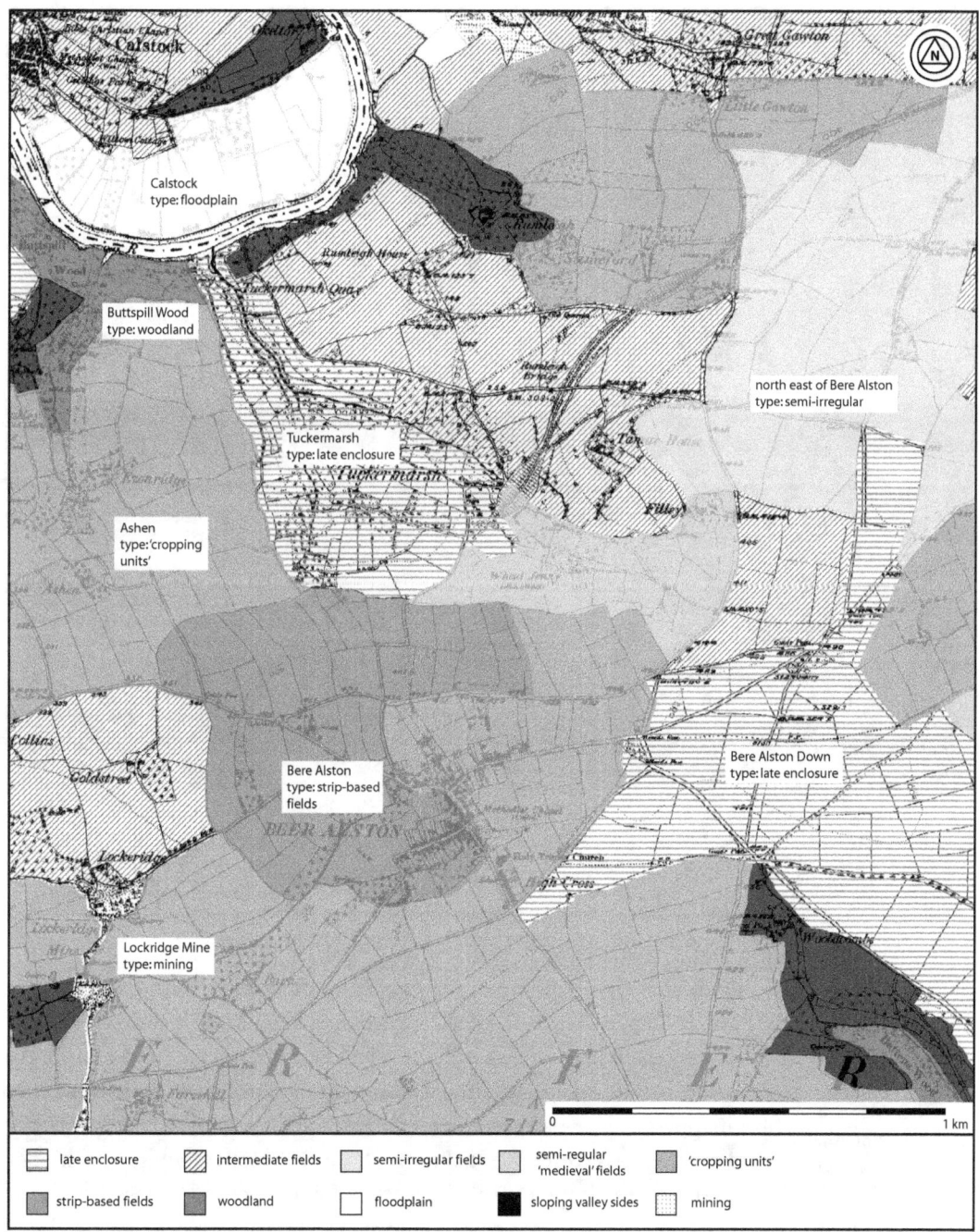

Figure 6.3: A characterisation of the historic landscape around Bere Alston. Based on the Ordnance Survey First Edition Six Inch map.

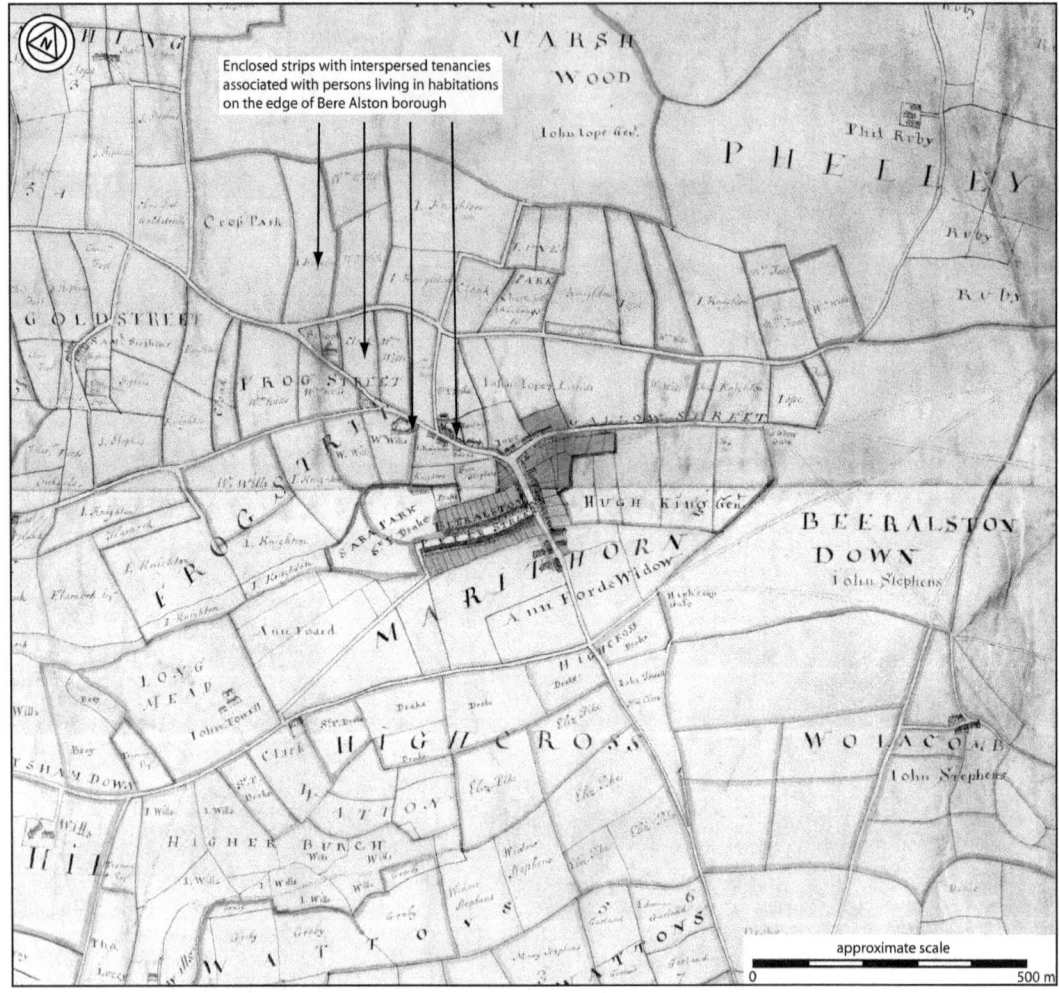

Figure 6.4: Extract from the 1737 estate map showing Bere Alston and surrounding areas (CRO ME2424). Note that 'Beeralston Down' had not yet been enclosed (cf. Figure 6.3).

of fields that clearly relate to the enclosure of open ground (such as Bere Alston Down, as described above: Figures 6.2–6.4), while what have been called 'intermediate' fields are slightly less regular, but still seemingly relatively recent in origin. When these different landscape character types are mapped a clear picture emerges in that the areas of earlier, probably medieval, enclosure concentrate on the lower slopes, while the higher ground remained open until the post-medieval period. Overall, there is nothing in the character of the landscape of Bere Ferrers that singles it out from the adjacent areas that did not experience mining history, with the possible exception of the small town at Bere Alston (see below).

For the parish of Bere Ferrers more detailed analysis was carried out on the Tithe map of 1845 (Figures 6.5–6.7).[7] This map was transcribed

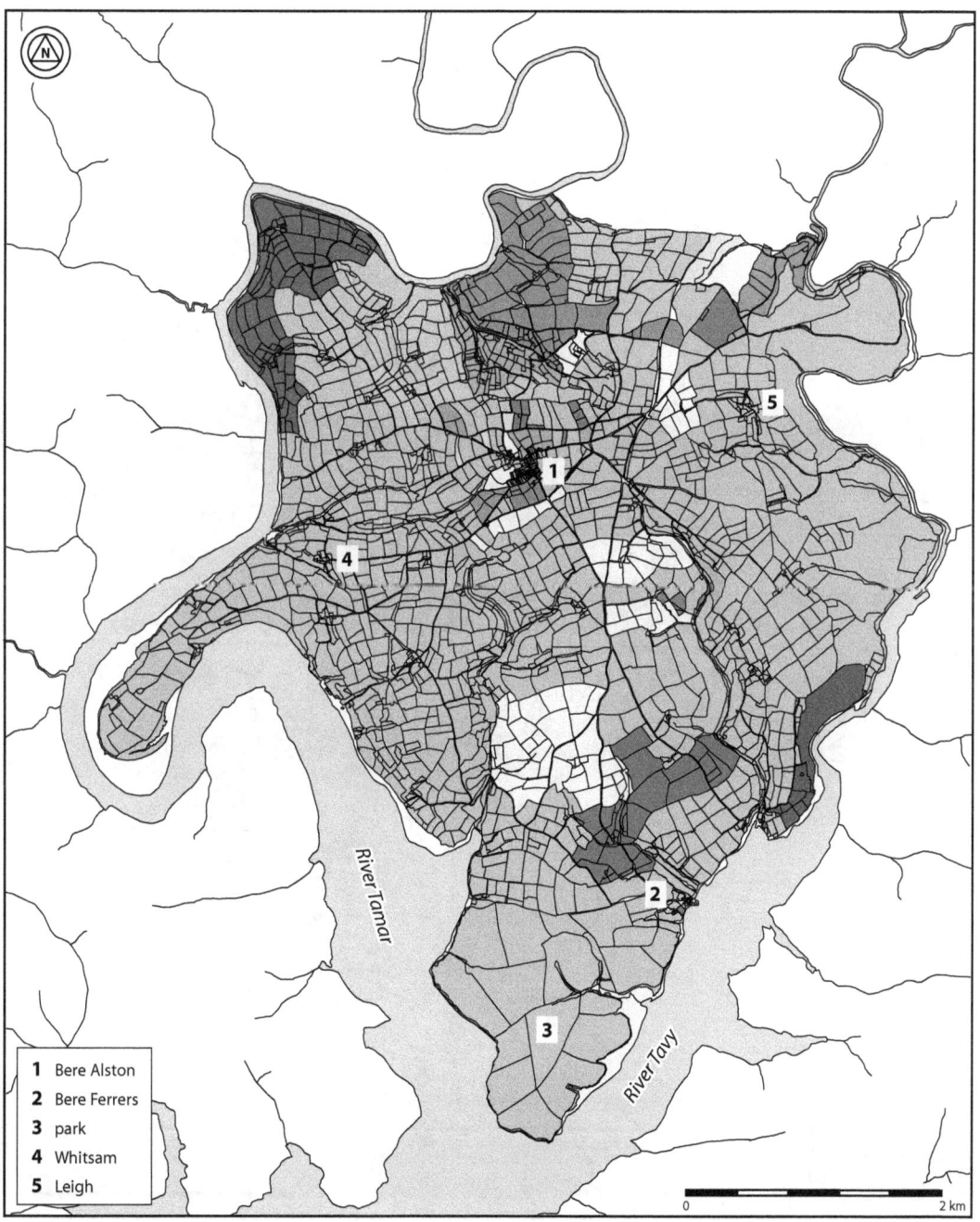

Figure 6.5: Bere Ferrers parish: land ownership as recorded in the Tithe survey. Note that while most of the parish was owned by the Earl of Mount Edgcumbe, there were scattered parcels to the north of Bere Alston (No. 1) in different hands (see Figure 6.12 for greater detail of this area).

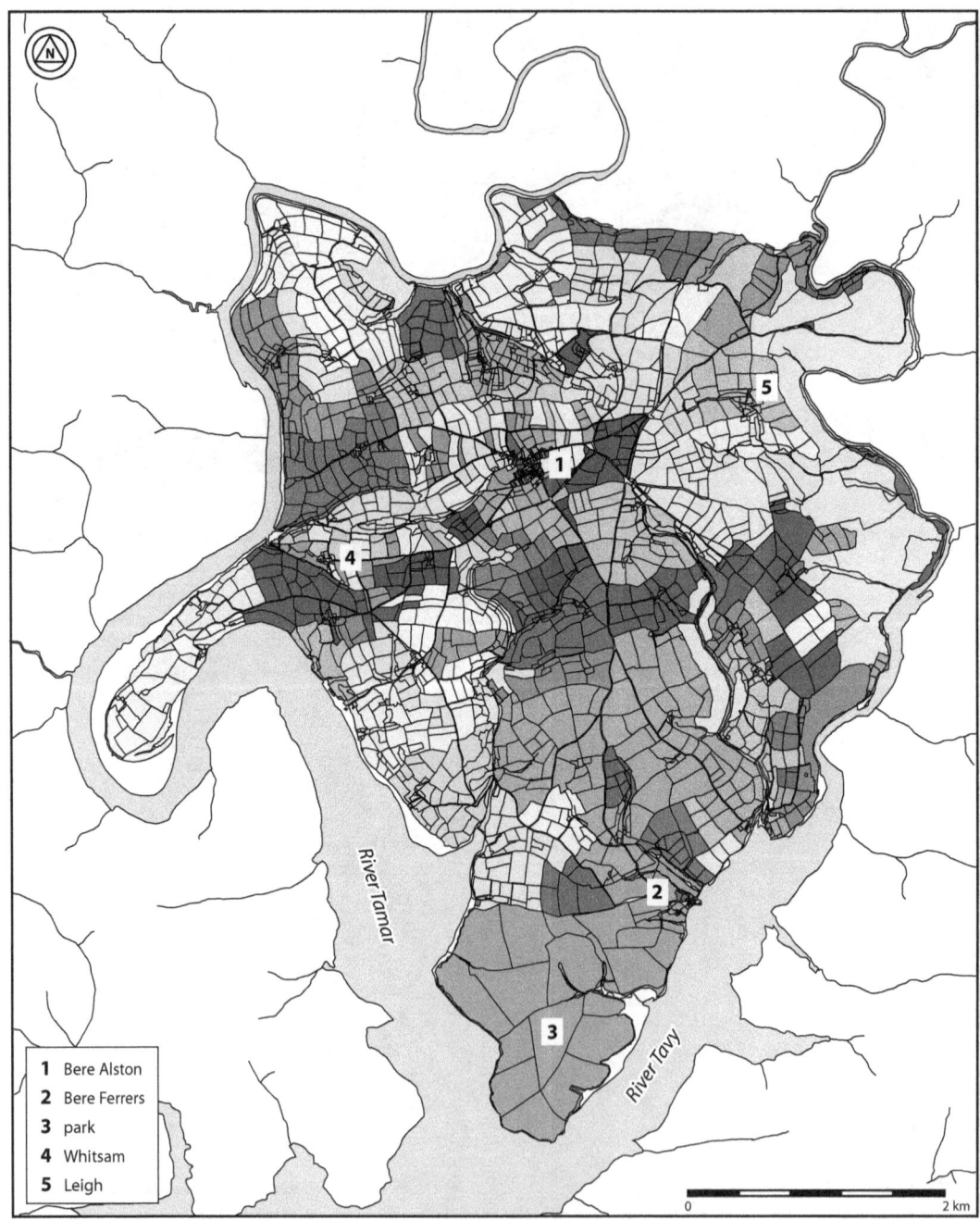

Figure 6.6: Bere Ferrers parish: land holding (i.e. tenements) as recorded in the Tithe survey. Most of the tenements on the Earl of Mount Edgcumbe's estate were compact blocks of fields associated with individual farms (e.g. No. 5: Leigh), but to the north of Bere Alston (No. 1) there was far greater fragmentation.

The mining community 133

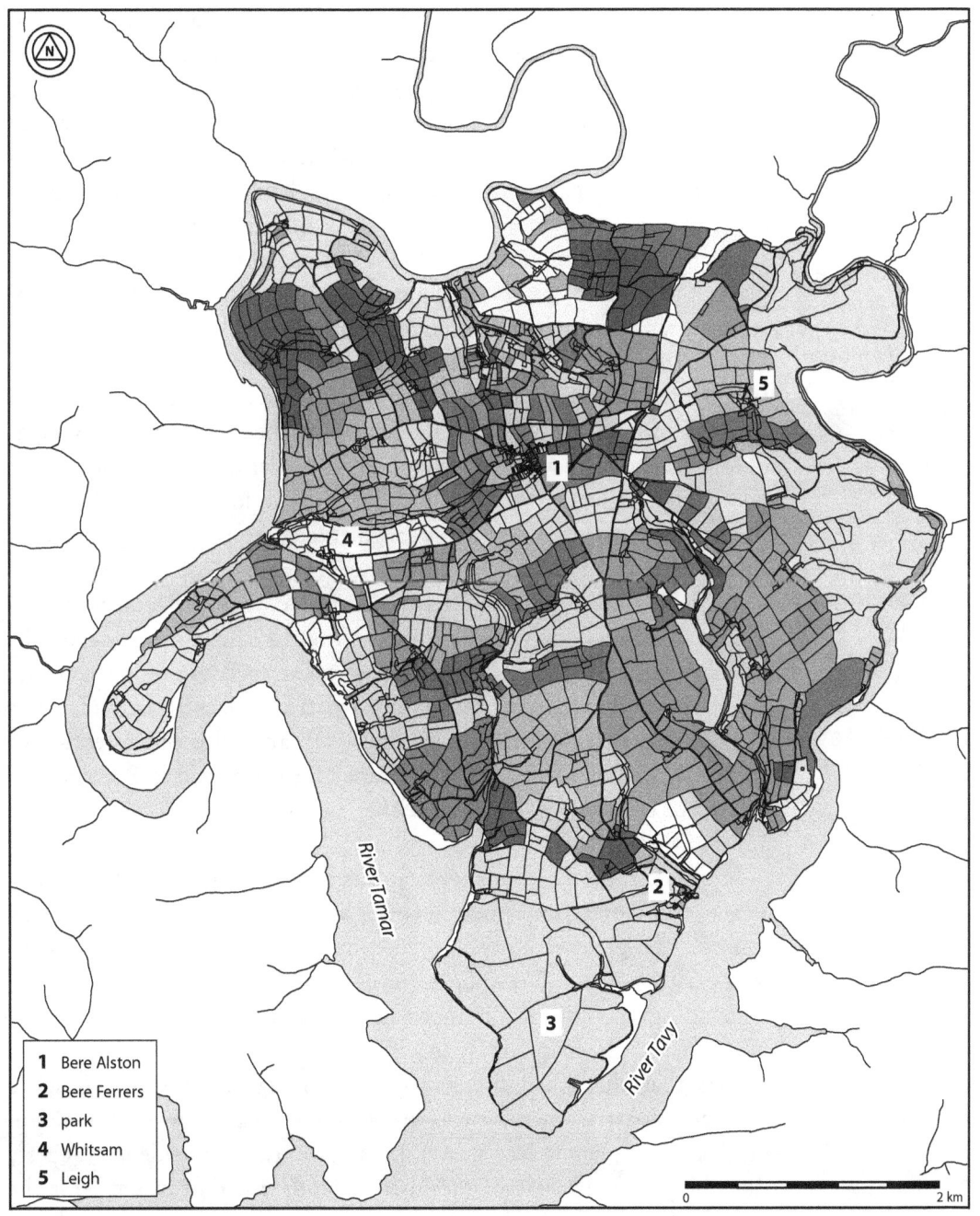

Figure 6.7: Bere Ferrers parish: land occupancy as recorded in the Tithe survey. Some tenants held several landholdings such as at Whitsam (No. 4), and to the west of Bere Ferrers hamlet (No. 2) and north of the former park (No. 3).

OPPOSITE

Figure 6.8: Characterisation of the settlement pattern in Bere Ferrers and adjacent parishes based on the Ordnance Survey First Edition Six Inch maps. This was a landscape typical of south-west England in that apart from the town at Tavistock, and the small towns at Bere Alston, Calstock, and Tamerton Foliot, there were few substantial nucleations, and most people lived in isolated farmsteads and small hamlets. While earthworks and the character of the historic landscape suggest some isolated farmsteads had once been hamlets, there is no evidence for parishes dominated by a single village.

into the GIS using the Ordnance Survey First Edition Six Inch maps as a base, with each parcel of land drawn as an individual 'polygon' (i.e. parcel of land) linked to a data-base which included information copied from the Tithe apportionment: the owner of the field, the tenement to which it belonged, the occupier, value, land-use, and field-name. It is important to map all three categories – landownership, tenements, and occupiers – as in places large areas of land were owned by the same individual but sub-divided into a series of discrete tenements. In places these tenements were themselves sub-divided between different tenants, while elsewhere the same tenant might hold several tenements. These various categories of Tithe data can be used to add further depth to the historic landscape characterisation, for example by looking at patterns of land ownership and occupancy: the interpretation of areas of long, narrow, curving fields around Bere Alston, for example, which on morphological grounds are suggestive of former open fields, is supported by their having a very fragmented pattern of land occupancy (suggesting that following the enclosure by agreement of these former open fields, tenements received the scattered strips of land that they had held immediately before enclosure). In contrast, most of the farmsteads outside Bere Alston village were surrounded by compact blocks of fields in single ownership, a pattern characteristic of closes held in severalty (Figures 6.5–6.7 and 6.12), and with the exception of Bere Alston there is little other evidence for open field in Bere Ferrers parish, in contrast to certain parishes further west such as Metherell and St Dominick (see below).

It is important to remember that this historic landscape characterisation is based on an interpretation of eighteenth- and nineteenth-century sources, and great care must be taken in back-projecting this evidence into earlier periods. We can, however, compare the landscape in 1845 – including the patterns of landownership and occupancy – with those portrayed on the map of 1737 (e.g. Figure 6.4, 6.5–6.7). This comparison reveals that there had been very little change in overall landscape character apart from the enclosure of some areas of former common on the higher ground (see above). In fact, a variety of evidence suggests that the physical fabric of the historic landscape dates to at least the later medieval period: a later fifteenth-century list of tenements within the parish lists the vast majority of the settlements depicted on the 1737 map (see below),[8] while the way that the mine workings and Lumburn Leat cut across what were clearly pre-existing roads and field systems proves they too are at least late medieval in date.

The mining community

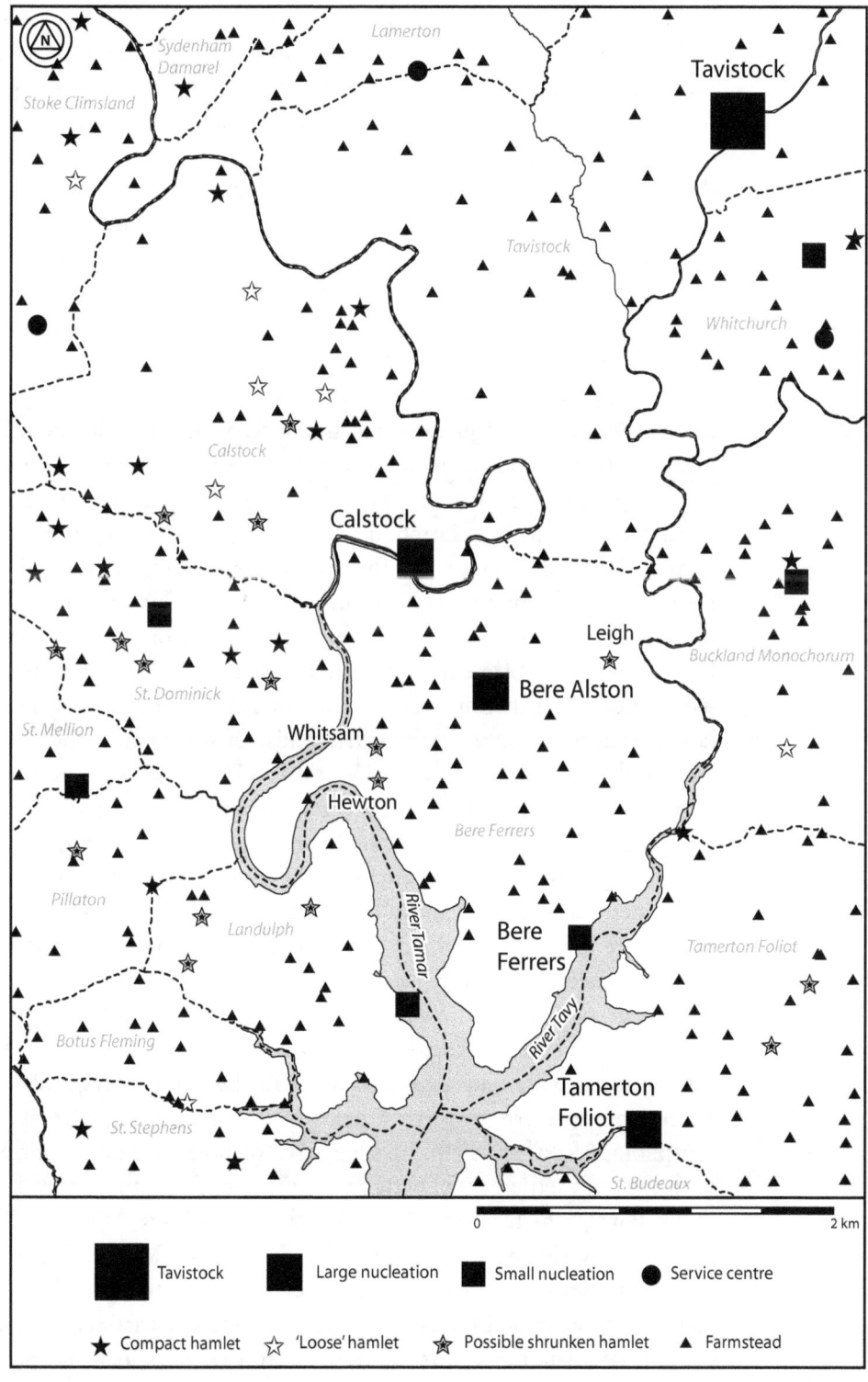

An additional layer of the historic landscape characterisation – an analysis of the settlement pattern – is shown in Figure 6.8. The nineteenth-century settlement pattern within Bere Ferrers parish was remarkable dispersed, with isolated farmsteads and small hamlets, each with their own field system, scattered across most of the parish, the exceptions being the far south, east and north, that appear to have been enclosed later. This settlement pattern is the same as that found elsewhere in south-west Devon (e.g. Tamerton Foliot, Buckland Monochorum, Whitchurch, and Tavistock in Figure 6.8) though in parts of eastern Cornwall there was a slightly higher number of hamlets (e.g. in Landulph, St Dominick and Calstock: Figure 6.8).[9] Elsewhere in the South West it can be shown that this settlement pattern replaced the late prehistoric/Romano-British landscape around the seventh to ninth centuries.[10] There were just two nucleated settlements in Bere Ferrers parish: Bere Ferrers itself, which in the nineteenth century consisted of the parish church, manor house (Bere Barton), school and a small cluster of cottages, and the town of Bere Alston with its church, three non-conformist chapels, an inn, a school, a cluster of farms and areas of terraced housing. To the east and north of Bere Alston the landscape is of a different character. The steeper slopes are cloaked in woodland, while the intervening areas of high ground have generally large, rectilinear, and straight-sided fields. In some cases, such as the area marked on the 1737 map as 'Beeralston Down', these were created between 1737 and 1845 through the enclosure of open pasture. The morphology of the fields around Newhouse is very similar, and the map of 1737 shows that the enclosure of this area was still underway; the names 'Great East Demean' and 'Heath East D.menes' suggest that these were areas of demesne pasture. The absence of Newhouse from the later fifteenth-century list of tenements within Bere Ferrers parish also suggests that it was established in the post-medieval period.

An area of altogether different character occupies the far south of Bere Ferrers, where an area of large, straight-sided, polygonal fields covers the southern tip of the peninsula (Figures 6.1 and 6.9). The morphology of these fields is suggestive of post-medieval enclosure, with the only settlement shown on the nineteenth-century maps being 'New Barn', which is not shown on the map of 1737. On the Tithe Map of 1845 this area was owned by the Earl of Mount Edgcumbe along with most of the parish (Figure 6.5), although the fields with this distinctive morphology formed a discrete tenement that also included the former manorial centre of Bere Barton (Figures 6.6 and 6.7). A

The mining community 137

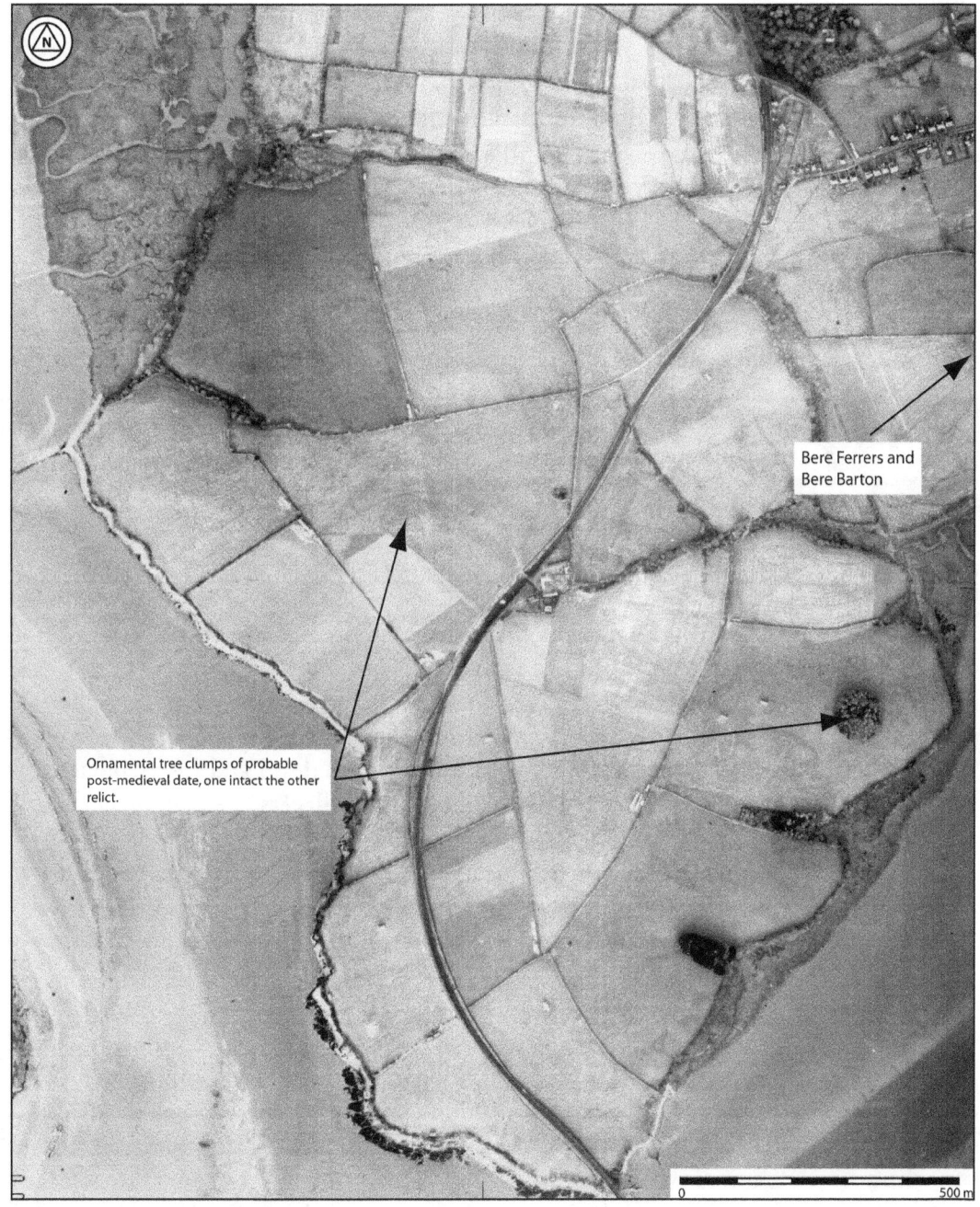

Figure 6.9: Vertical aerial photograph taken in 1946 of the former medieval park in the far south of Bere Ferrers parish (CPE/UK/1890: 10DEC '46 F20" // MULTI (4) 54 SQDN. Frame 2317).

map of the Plymouth area dating to some time before 1549[11] labels this area as 'Beere Parke', while a later sketch-map of Bere Ferrers parish, probably seventeenth century in date,[12] specifically records it as 'The Park or Mr Shepards Lands'. On the same map, Mr Shepard is shown as the owner of Bere Barton. While the name 'Park' is a common element of Devon field-names, deriving from the Old English *pearroc*, meaning 'an enclosed piece of land',[13] the specific labelling of this area as Parke or Park on these early maps, and the tenurial link with the manorial centre, suggests that this was a medieval deer park. This is further supported by John Leland's description of the area: 'Then is the uppermost where Tave Water cummith onto Tamar. And on the east side of this creek is Buckland. And on the west side is Bere where the Lord Brokes house and park was.'[14] Lord Robert Willoughby de Broke died in 1552 in Bere Barton, and his tomb remains in St Andrew's church.

The Domesday landscape

The earliest description of what we now know as Bere Ferrers is the Domesday record of the manor of *Birland*:

> Reginald holds Birland from the count [of Mortain]. Ordwulf held it before 1066. It paid tax for 4 hides. Land for 15 ploughs. Reginald has 1 hide in lordship. 16 villagers, 5 smallholders and 5 slaves who have 6 ploughs and 3 hides. 3 pigmen who pay 15 pigs. 7 salt-houses which pay 10s. Pasture 5 furlongs long and 1 furlong wide; woodland 1½ leagues long and 1 furlong wide. 1 cob; 5 cattle; 3 pigs; 30 goats. Formerly 60s; value now 100s.[15]

Gover *et al.* describe *Birland* as a place-name of unknown meaning,[16] though it may be derived from britonic *byr* or *ber*, meaning a pointed piece of land or promontory (as in the Beara Peninsula in West Cork).[17]

Later sources show that the manorial centre of *Birland* was at Bere Ferrers, with the site of Bere Barton being the seat of the manor since at least the fourteenth century. In interpreting the Domesday description one must always remember the fallibilities of this survey, and in this case the number of ploughteams in lordship are not listed; the absence of sheep is also curious as most Devon manors possessed some. The population seems a little low for a manor this size, assuming

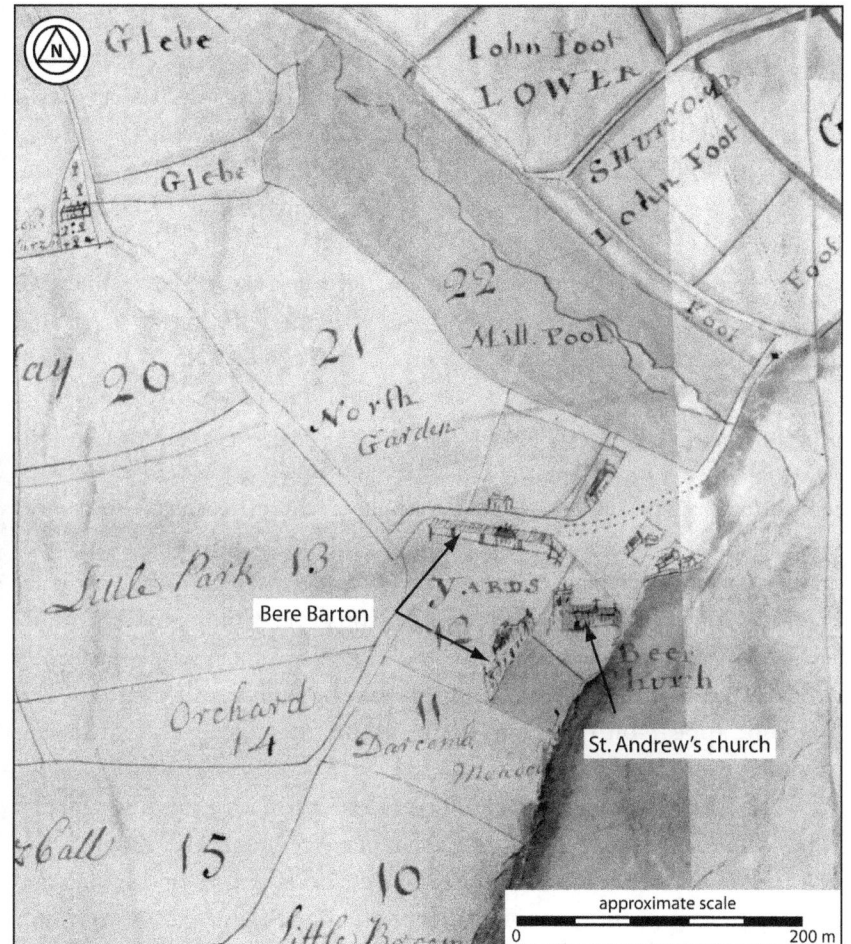

Figure 6.10: Extract from the 1737 estate map showing the small hamlet at Bere Ferrers, including St Andrew's church and the manor house at Bere Barton (CRO ME2424). The 'Mill Pool' relates to the 'Salt Mill', a tide mill that existed up until at least the late seventeenth century.

that it was coterminous with the later parish which covered the entire peninsula, though in part this may be explained by the abundance of woodland that presumably cloaked the steeper slopes, and, large area of open pasture that may have occupied the higher ground north and east of Bere Aston. Given the location of the manor, bounded on either side by the tidal shores of the rivers Tamar and Tavy, it is unsurprising that there was a salt industry.

The manorial centre at Bere Ferrers

The map of 1737 shows that the historic core of Bere Ferrers, comprising St Andrew's church, the manor house of Bere Barton, an inn, and a small number of roadside cottages, lay on the western bank of the river Tavy beside 'Mill Pool' (Figure 6.10). The fields that surround the hamlet are predominantly large, of rectangular to sub-rectangular shape, and show none of the characteristics of former strip-based open field cultivation that is seen around Bere Alston. The pattern of landownership and occupancy in 1737 and 1845 – a series of compact blocks of fields in the same hands – is also suggestive of closes held in severalty (Figures 6.5–6.7).

Henry de Ferrers held the manor of Bere Ferrers in the second half of the twelfth century,[18] and in 1337 Sir William de Ferrers was granted a licence to crenellate his manor house there.[19] This licence was not unusual and there were almost twenty such grants awarded in Devon during this period.[20] It has been claimed that Bere Barton contains some surviving medieval architecture including a section of castellated wall abutting the rear of the main house and connecting it to a large barn,[21] but fieldwork by the authors suggests that this 'castellated wall' was formerly part of the adjacent barn which at some point has been shortened, and that the 'crenellations' are simply the truncated remnants of a series of window reveals. This barn was apparently constructed within the middle years of the eighteenth century as it does not appear on the Mount Edgcumbe estate map of 1737, but is included on a map of 1784.

While it is possible that Bere Ferrers had a role as a landing place, it is not mentioned as such in the Exchequer Accounts, although a series of quays are documented at other places along the rivers Tamar and Tavy which were more convenient for the mines, sources of wood for fuel and timber, and the smelting sites (see Chapter 4). Henry[22] alludes to a local tradition that the road running though Bere Ferrers and down to the Tavy was called 'Silver Street' because it was the route via which silver from the medieval silver mines was transported. The name 'Silver Street', however, occurs in various places elsewhere which have no connection with either silver or mining, for example at Braunton in North Devon,[23] and Congresbury in Somerset,[24] suggesting that this name has other origins. In fact, there is no evidence that the settlement at Bere Ferrers played any role in the silver mining industry, and indeed, as silver was a Crown prerogative, the opening of the mines would have not given the lord of the soil a direct source of income. The establishment of the borough at Bere Alston in 1305,[25] with a market each Wednesday, would, however, have provided the lord of the manor with revenue, while the rebuilding of the church was also paid for by tithe revenue from the produce of the mine, with the tenth dish of ore being assigned to the church and bought back by the Crown at two shillings per load.[26]

The borough at Bere Alston (Figure 6.2–6.4)

In 1295 Reginald de Ferrariis, lord of the manor of Bere Ferrers, was granted a weekly market and an annual fair[27] and by 1305 the lord was receiving a one shilling rent from each of twenty burgesses.[28] The location of this borough was at the core of the modern settlement of Bere Alston and represents an attempt by the lord of the manor to exploit the opportunities for raising revenue afforded by the developing mining industry through the rents paid by burgesses in return for being exempt from any market tolls. The miners were also granted exemption as one of their privileges.[29] On the other hand, local producers and strangers bringing in produce would have contributed, through the tolls they paid, to the revenues of the borough and in this way the lord of the manor was able to profit indirectly from the increased population and demand for food and other supplies brought about by the opening of the mines. Lying just one kilometre from Lockridge Hill, the borough at Bere Alston benefited from servicing the mining industry, but its success was clearly not wholly dependent on the mines as it survived as a prosperous agricultural market until at least 1600, by which time there is little documentary or archaeological evidence for mining within the parish (it was not until the eighteenth century that it resumed on a significant scale).[30]

The foundation of Bere Alston was not, however, an unusual event in Devon as the county had a total of 74 boroughs, the greatest number of any county in England, almost half of which were founded in the late thirteenth or fourteenth century.[31] The borough at South Zeal to the north of Dartmoor, for example, was also established under seigneurial authority at the turn of the fourteenth century and was of comparable size, with twenty burgesses paying rent in 1315 (Figure 6.11).[32] This planned town was created within South Tawton parish, through which the Exeter to Okehampton road passes, and it was either side of this major highway that the planned new town was created. The neighbouring county of Cornwall ranks third in the country in terms of the number of boroughs, and both Devon and Cornwall retained these positions throughout the medieval period.[33] Beresford and Finberg[34] suggest that this abundance of 'petty boroughs' was due to the relatively late development of rural settlement in largely

Figure 6.11: The planned medieval borough at South Zeal. Based on the Ordnance Survey First Edition Six Inch map.

unsettled areas, a situation dictated in part by the terrain of these counties, but this view of the South West landscape as being one of late colonisation is no longer sustainable. The density of population and ploughteams in Domesday is comparable to the rest of southern England, and palaeoenvironmental sequences consistently show that the region's lowland areas were extensively cleared of woodland and settled from late prehistory onwards.[35] More likely explanations for the relatively high density of boroughs in Devon is the highly diversified economy of the region and the fragmented nature of lordship: where a single lord held all the manors in an area then it was usual for only one borough to be founded, whereas if manors were in separate hands there was a tendency for each lord to want to found their own.[36]

The earliest known plan of Bere Alston is the estate map of 1737 (Figure 6.4), although this requires careful interpretation as it is a little schematic in places necessitating the use of later maps – notably the Tithe Map of 1845 – in the analysis of its plan (Figure 6.12). The settlement plan can be divided into four units. To the north-west

Figure 6.12: Patterns of landownership around Bere Alston recorded in the Tithe survey of 1845. Note how the very fragmented tenements to the north of the settlement correspond to fields whose long, curving shape suggests they were derived from the enclosure of strips in a former open field.

there was a cluster of four farms (which in 1737 were held by Messrs Wills, Knighton, Jope and Cloak), several cottages, and a pound spread along both sides of Frog Street, that appears to curve around what may have been a small oval-shaped green. These buildings lay within small, irregularly shaped plots of land (tofts), to the north and west of which the field boundary pattern is clearly derived from the enclosure of former open field, with a distinctive pattern of long curving fields laid out between a series of sinuous boundaries which presumably represent consolidated groups of former strips within furlong blocks. The fields held by the four farms of Messrs Wills, Knighton, Jope, and Cloak were scattered right across this area, being intermingled with each other, and together embracing virtually the closes within the putative former open field. Field-names in this same area such as Bovetown, and Lower, Higher, and Outer Bove Town, are also indicative of common field agriculture.[37]

To the south of the farms along Frog Street, the 1737 map shows that there were three blocks of terraced housing spread along the north–south oriented Fore Street (on the eastern side of which lay the chapel of the Holy Trinity built in the mid-fourteenth century[38]), Gallow Street (now Bedford Street) which ran east from the top of Fore Street towards 'Beeralston Down', and Pepper Street (now Cornwall Street) which ran west from the bottom of Fore Street towards Lockridge Hill. The map of 1737 shows Pepper Street running down the centre of what is shown as a large rectangular block of burgage-like plots (Figure 6.4), and the impression is that this is the core of the medieval borough, very much like the planned town at South Zeal (Figure 6.11). A road that runs from the western end of Pepper Street to Furzehill clearly cuts across an earlier, probably open, field system, and it is likely that this route served to connect the market at Bere Alston with the mines south of Furzehill and the documented quay at Lockridge Pill, referred to as 'Luxeruggepulle' in 1306.[39] This relationship with the mine complex presumably explains why the borough was laid out along this new street as opposed to the existing north–south road between Frog Street and Bere Ferrers.

Overall, it is tempting to see Frog Street as a pre-borough hamlet associated with a small open field, to which the town was later added with one block of tenements either side of Pepper Street and a possible second block either side of Gallow Street. A key question is whether the apparent regularity in the layout of Bere Alston was due to it having been laid out as a planned town, or because the tenements were added gradually, re-using former open-field strips. The Pepper Street and Gallow Street blocks of tenement plots appear to have been laid out between three parallel but curving boundaries, each c.350 feet (c.100 m) apart, which have the appearance of furlong blocks. This curvature, however, is probably due to the southern boundary being a natural stream, and the others having been laid out parallel to it. The tenements within these Pepper Street and Gallow Street blocks are straight-sided as opposed to the curving form that would be expected if these had been former open-field strips. The distances between the three parallel boundaries, c.330–360 feet [100–110 m] is also rather short compared to the furlong blocks around Bere Alston, and indeed, Metherell and St Dominick, which are typically 500–600 feet [150–180 m] wide. Finally, the configuration of the curving boundaries of two large closes to the north-east of the Pepper Street tenements, held by Sara Park in 1737 (Figure 6.4), bears no relationship to the Pepper Street block. There are also a series of faint earthworks on aerial photographs that are parallel to these curving field boundaries, which, taken together, are suggestive of a series of strips that appear to lie unconformably beneath the Pepper Street block (Figure 6.2).

It would appear, therefore, that the burgage plots at Bere Alston were superimposed upon an earlier open-field system. The southern block of tenements, either side of Pepper Street, appears to have been the main focus of the settlement, while the northern block, either side of Gallow Street, appears to have been partly deserted (if it was ever successfully occupied). It is possible that the open fields which they were superimposed upon had already been enclosed, as at Tavistock, just ten kilometres north-east of Bere Alston, where the amalgamation and purchase of open-field strips, and their conversion into closes held in severalty and enclosed with hedgebanks, was well advanced by the first quarter of the fourteenth century.[40] Alternatively, it may have been that the emergence of the town, along with the increased local demand for food including that from the mining community, would have provided local farmers with a regular point of sale and a captive market for their produce, in the same way that Herring suggests that Cornwall's tin industry stimulated agricultural intensification and the consolidation of communal field systems.[41]

Farmsteads and their field systems

Apart from the hamlet at Bere Ferrers, and the town at Bere Alston, the settlement pattern in Bere Ferrers parish in both 1737 and 1845 was characterised by isolated farmsteads and small hamlets of just two or three tenements (Figures 6.8 and 6.13). Although none of the medieval settlements within Bere Ferrers has seen any archaeological investigation, fieldwork on Buckland Down (also known as Roborough or West Down), within the neighbouring parish of Buckland Monochorum, tells us something of their likely character.[42] Four sites have been

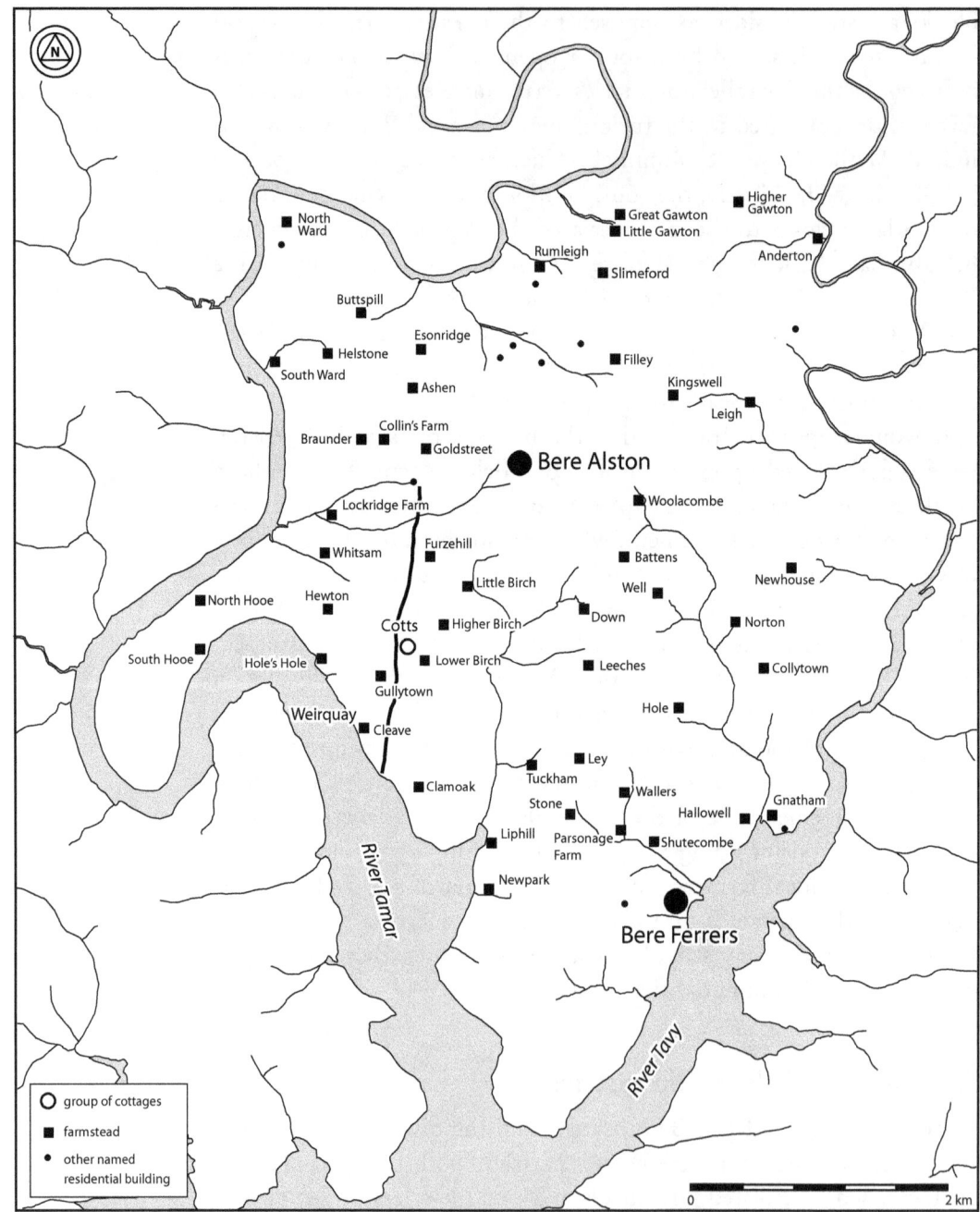

Figure 6.13: Nineteenth-century settlements in Bere Ferrers parish (from the Ordnance Survey First Edition Six Inch Map).

investigated, each composed of earthworks suggestive of a longhouse surrounded by ancillary buildings and surrounded by a field system. It has been suggested that the exploitation of Buckland Down may have been linked with the foundation of Buckland Abbey in 1278, and which in the fourteenth century was granted rights to clear, enclose, improve, and cultivate parts of the Down.[43]

While page 136 in Dean and Chapter of Exeter Cathedral archives MS 3522 (see below) shows that most of the settlements in Bere Ferrers parish in 1737 existed in the late fifteenth century, and other documentary sources show that most settlements can be traced back at least as far as the early fourteenth century, only the tenement called Leigh, first referred to in 1284,[44] is recorded before the opening of the mines, although this is simply due to the absence of earlier documentation for the manor. There is some evidence from Bere Ferrers parish, however, that what by the eighteenth and nineteenth centuries were small hamlets of just two or three tenements, or even isolated farmsteads, had previously been somewhat larger. In his parochial history, Beddow[45] refers to an eighteenth-century gamekeeper's cottage as well as seven other dwellings at Whitsam, for example, though these have now been mostly demolished or incorporated into new structures. The representations of both Hewton and Leigh on the 1737 map are similar in scale to that of Whitsam so it is possible that there, too, numerous cottages might have existed that have since been demolished. This process of settlement shrinkage is not unique to this area, and examples of what were small hamlets during the High Middle Ages becoming isolated farms by the eighteenth and nineteenth centuries are known across Devon and Cornwall.[46]

These settlements were associated with compact blocks of fields that are in sharp contrast to the intermingled landholdings seen around Bere Alston. The morphology of the fields beyond the immediate vicinity of Bere Alston is also suggestive of closes held in severalty. In the west of the parish the landscape is characterised by small, broadly rectangular fields, some with a more irregular shape, whereas the fields on a series of farms to the south and east tend to be slightly larger and more regular in their layout (e.g. Ley, Leeches, and Battens). Most of these field systems have a distinctive morphology, being broadly rectilinear, but often with slightly curving sides and laid out between long, sinuous, and roughly parallel boundaries. These are reminiscent of what have been called 'cropping units' in previous Historic Landscape Characterisations of Cornwall and Devon, within which it has been

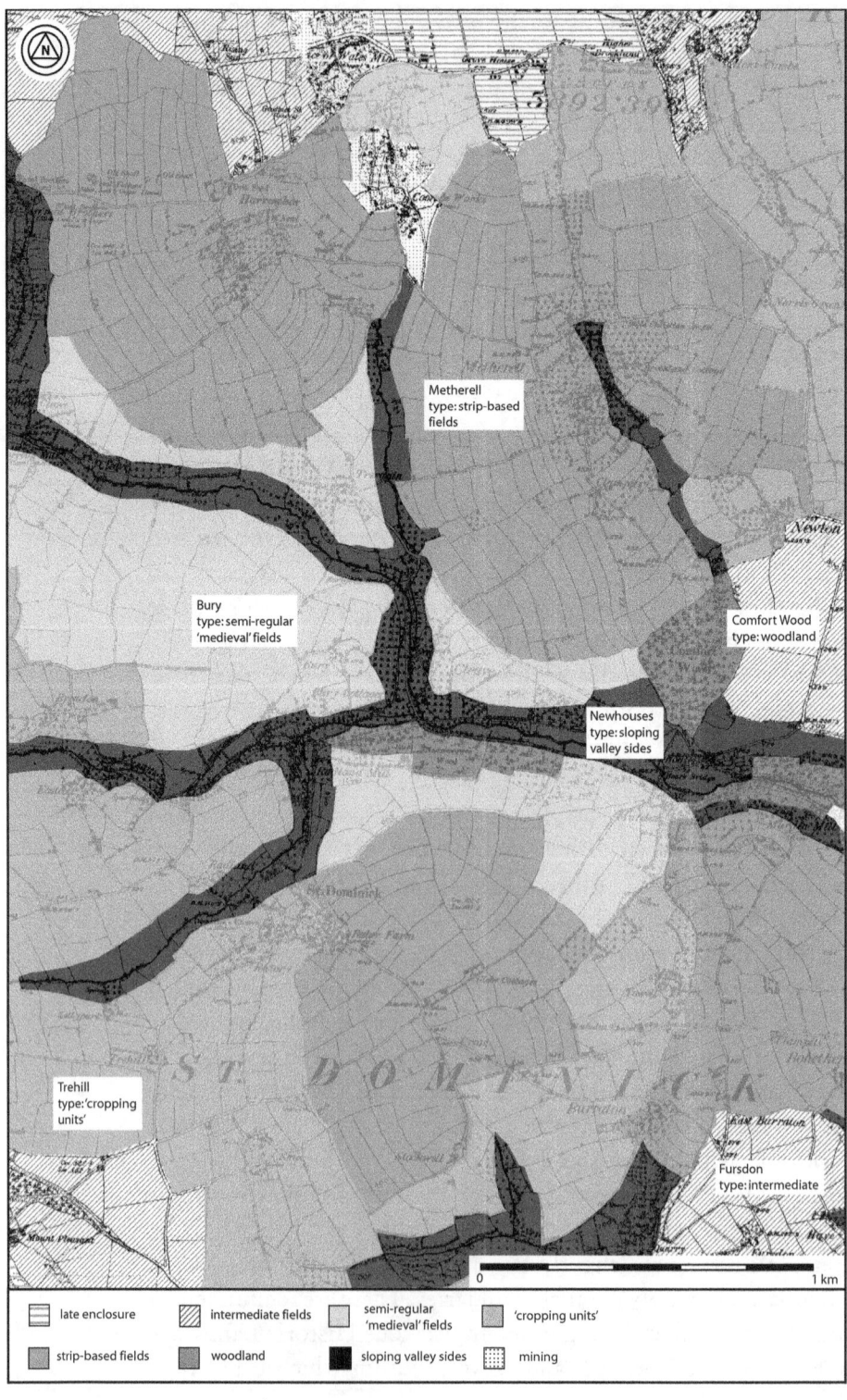

suggested that each field represents a consolidated block of former open-field strips.⁴⁷ Around Metherell and St Dominick, to the west of the Tamar, the evidence for former open field is clear (Figure 6.14), whereas across most of Bere Ferrers there are so few long, narrow fields that this interpretation may not apply: rather than consolidated blocks of former open-field strips, some of these 'cropping units' may simply be closes held in severalty, albeit with curving sides reflecting their use for arable cultivation (e.g. Figures 6.15–6.16). The one area of Bere Ferrers parish where there may have been open field is around Bere Alston where there are a series of furlong-like blocks of long, narrow, fields whose interpretation as former open-field strips is strengthened by the highly fragmented pattern of eighteenth- and nineteenth-century landownership and occupancy (see above).

The antiquity of these field systems in the west of Bere Ferrers is difficult to establish, though the small open-field system around Bere Alston appears to pre-date the foundation of the borough in the early fourteenth century. To the south-west, the field systems – both in terms of their physical boundaries and patterns of land occupancy – appear to be cut by the line of mine workings dating to 1292 and later (Figure 6.16). This confirms the logical assumption that mining was superimposed upon a pre-existing agricultural landscape, and in about 1300 Reginald de Ferrers, lord of the manor, petitioned the Crown to get damage to his land caused by the mines and miners remedied by the Crown.⁴⁸ In a separate petition, undated, but also thought to be of c.1300, 'Reynald' (Reginald?) de Ferrers requested compensation at the rate given in the Peak District mining areas, which was equal to one thirteenth of the ore, for damage to his land and to his people.⁴⁹ There is no evidence to suggest that these, or similar petitions, were successful and while they illustrate the negative impact of mining on the existing landscape, they may also represent an attempt by Reginald de Ferrers as lord of the manor to get at least some financial reward from this Crown prerogative.

Large areas in the western half of the parish appear to have been unenclosed until the post-medieval period. Areas such as 'Beeralston Down' were still unenclosed in 1737, while 600 acres of Morwell Down to the north, between Bere Alston and Tavistock, was not enclosed until 1836.⁵⁰ It is possible, however, that these unenclosed pastures were periodically cultivated, since evidence clearly demonstrates this practice in nearby Tavistock. In 1310, for example, two and a half 'ferlings' at Newton were granted with common pasture for all the

OPPOSITE

Figure 6.14: A characterisation of the historic landscape around Metherell and St Dominick, based on the Ordnance Survey First Edition Six Inch Map. Note how the enclosed blocks of former strip fields, comparable to those north of Bere Alston (Figure 6.12), are somewhat different in character to those at Whitsam (Figure 6.15).

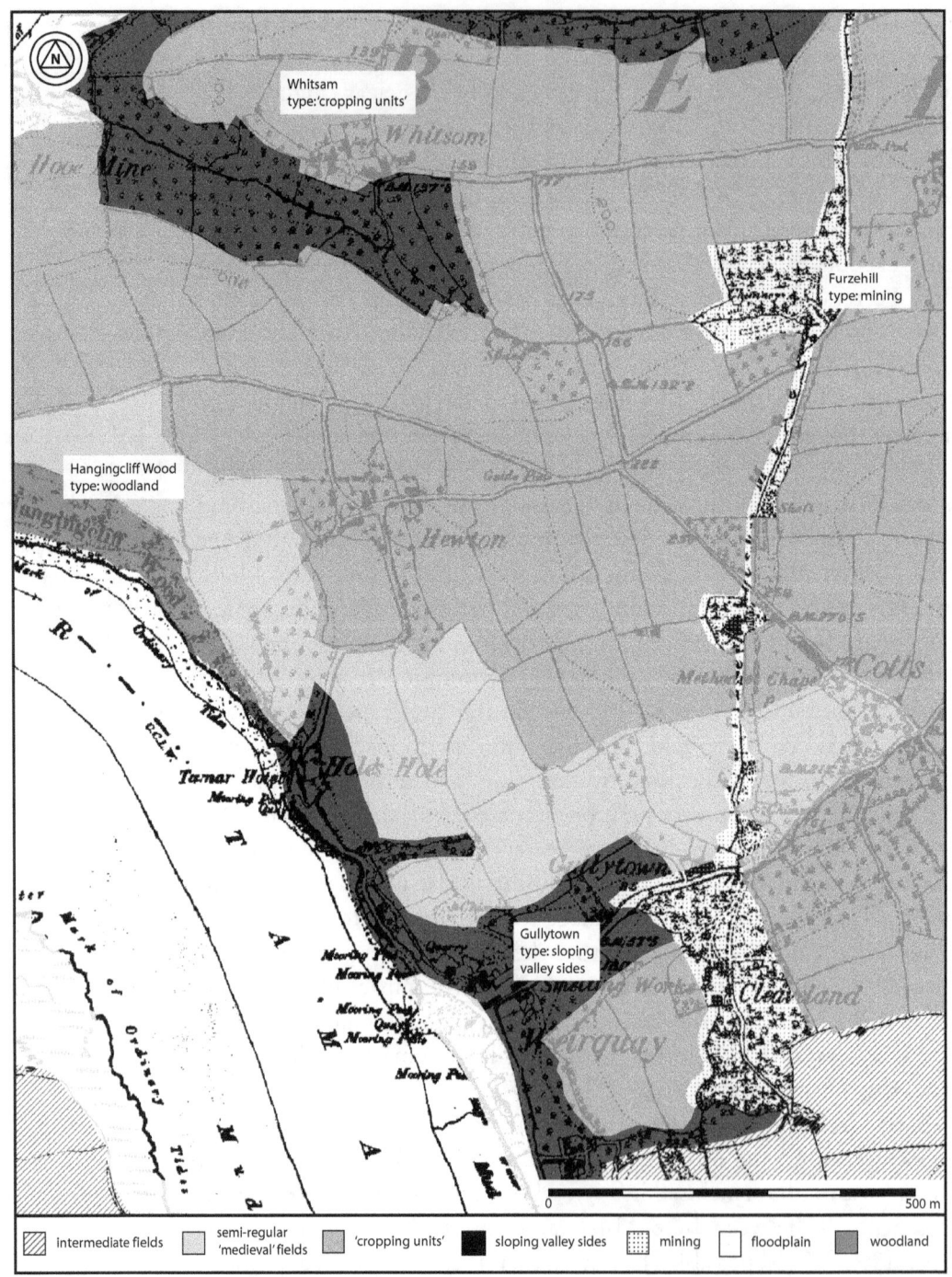

Figure 6.15: A characterisation of the historic landscape around Whitsam, based on the Ordnance Survey First Edition Six Inch Map. Note how the cropping-unit type fields are somewhat different in character to the blocks of former strip fields around Metherell and St Dominick (Figure 6.14).

Figure 6.16: Extract from the 1737 estate maps of the area around Whitsam, showing a landscape overlain by the mine workings. See Figure 6.15 for a characterisation of the Ordnance Survey First Edition Six Inch map this same area.

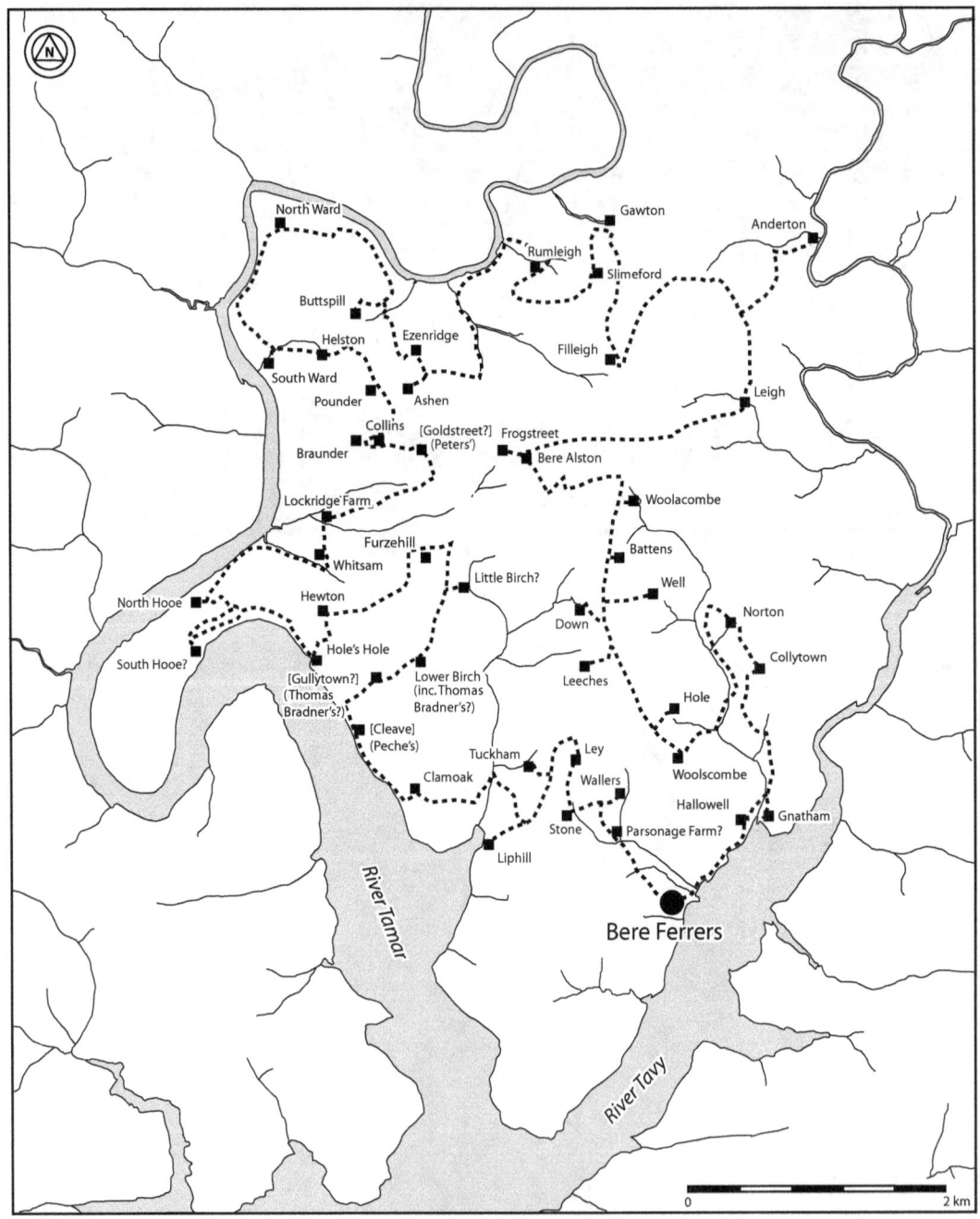

Figure 6.17: Places in Bere Ferrers parish referred to in the late fifteenth-century 'confessor's itinerary' (after Henry, 'Silver and salvation').

grantee's cattle that could winter upon Luscombe Down [north of Morwell Down] 'when the said waste lies untilled and not let off'.[51] The Bere Ferrers estate map of 1737 and Tithe Map of 1845 both show that the steep slopes in the eastern part of the parish were heavily wooded, as they are today. Documents relating to the silver mines do refer to woodland in these areas being exploited for fuel and timber (Chapter 5), and the fact that smelting activity was moved to Calstock on the Cornish side of the river Tamar, to exploit the king's woods in that parish, suggests that woodland in the eastern half of Bere Ferrers, as well as the Abbot of Buckland's woods at 'Biccombe' (Bickham), was depleted.

A key document in establishing the antiquity of the settlement pattern is page 136 in MS 3522 of the Dean and Chapter of Exeter Cathedral archives. This is a fifteenth-century list of places within Bere Ferrers parish, which shows that the vast majority of the farms depicted on the 1737 map existed at this time (Figure 6.17). The purpose of the document is, however, disputed. Henry has suggested that it describes the route to be followed by a priest hearing confession,[52] although Orme rejects this on the grounds that in medieval England confession was always heard in church. Instead, he proposes that the memorandum represents a rota for the occupants of each settlement to attend the parish church and give confession.[53] In response, Henry[54] has rejected this on the grounds that the memorandum recites a logical topographic route for an itinerary, and that these settlements would not have been able to function if left empty for a day. Although Orme has made no further response, it can be observed with regard to these final arguments of Henry that what is a logical route for an itinerary is also a logical sequence for a rota of attendance, and that the short journey to the church at Bere Ferrers would not have meant leaving tenements deserted for a whole day. Henry also suggests that reference to tenements using personal names (Peche's, Thomas Bradner's, and Peter's) was an attempt to distinguish between the occupants of clusters of buildings, which would be necessary if the priest was visiting these places, although this can be rejected on the grounds that it was quite common for late medieval tenements to be given personal names.[55] Overall, the balance of the argument appears to be in favour of Orme's interpretation of this document as the order in which this scattered population went to church to have their confessions heard, rather than a priest's itinerary. Either way, the list allocates an amount of time to each place which is presumably in proportion to the size

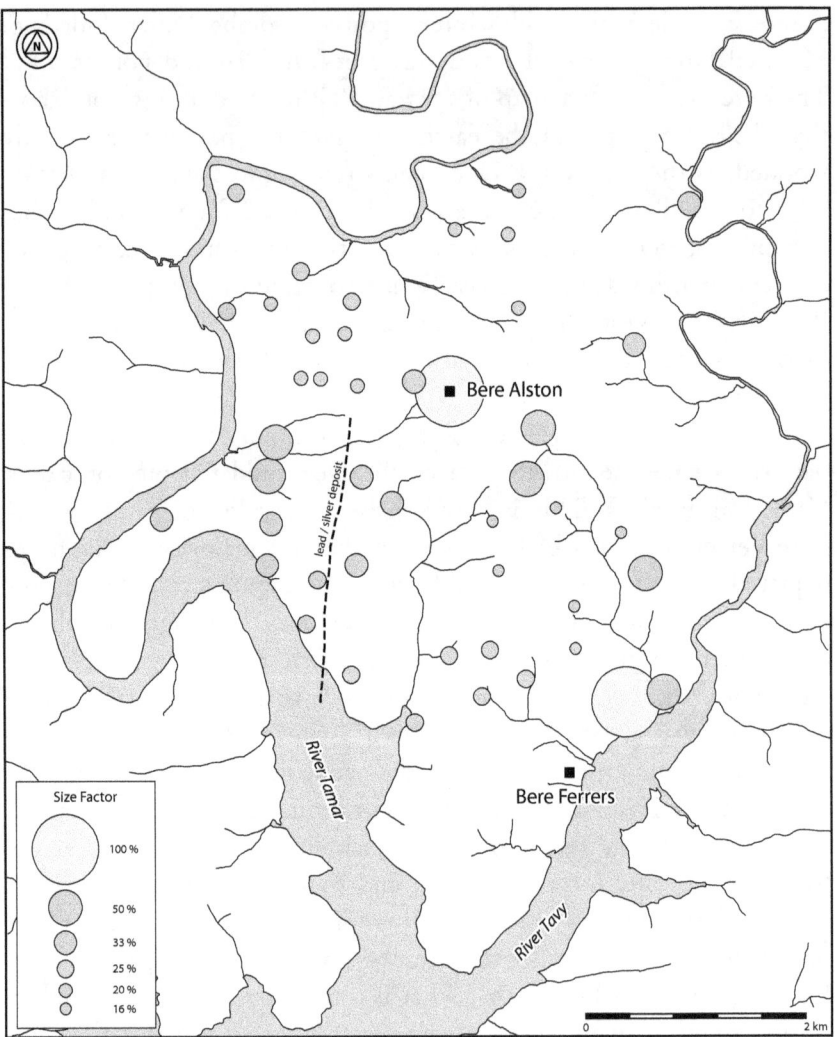

Figure 6.18: The relative size of settlements in Bere Ferrers parish based on Exeter Dean and Chapter MS3522, p. 136: assuming that it describes a list for attendance at the church (ORME, *Confession*) rather than an itinerary (after Claughton, *Silver Mining*, fig. 28). For a full discussion on their relative size and the apparently large settlement immediately north east of Bere Ferrers see ibid, appendix 12.

of the population. This suggests the largest population was at Bere Alston, but that settlements in the rural landscape closest to the mines had similar or slightly larger populations compared to those further away; there is no evidence from this that significant numbers of miners were housed in otherwise agricultural settlements adjacent to the mines (Figure 6.18).

This leads to another aspect of the interpretation of this document that is of greater relevance to this project. Henry argues that the visits were, or were intended to be, made to members of the mining community, rather than all parishioners, which implies that the miners lived in all of the agricultural settlements of the parish. Two distinct classes of people are referred to in the memorandum: *parochianorum* (parishioners), and *manentes* which is usually taken to mean either a rural land-worker, a servile person, the inhabitant of a parish or town, or the place inhabited by one of these people. She suggests that the latter group included the mining-related population, and that in this document the term means 'those staying at'. Her reading implies that miners and related workers were housed at nearly all of the settlements of the parish at this time, and proportionally more time was indeed allocated to settlements nearest the mines and quays, as well as at Bere Alston, suggesting that there were larger populations at these places. If Henry is correct, then this spreading of the mining population across the wider landscape might suggest dual-occupancy (i.e. part-time miners who were also engaged in agricultural activities), but her interpretation is far from convincing. If Orme's more straightforward view is correct, and *manentes* was used in its normal sense of a rural land-worker, a servile person, or the inhabitant of a parish or town, then there is no reason why the occupants of these settlements need have been involved in the mining industry. Following Orme's interpretation, the slightly larger size of the settlements in the vicinity of the mines may simply reflect the greater agricultural productivity of this area, and it is indeed the case that historic landscape characterisation indicates that the west of the parish contained the more ancient enclosures, which presumably correspond to the best agricultural land. To the east, in contrast, there were extensive areas of land that remained unenclosed until the post-medieval period, and were therefore presumably of poorer quality.

Documentary sources, principally wage and account rolls, include references to the payments made to workers for carrying out specific tasks, and many of their surnames relate to local settlements. These duties included transporting ore (Gilbert de Clyue [Cleave])[56] and fuel (Alice de ffeylegh [Filley] who was paid for four days carriage of fuel to the furnace by her horse),[57] for use of a horse (William de Gouetoun [Gawton]),[58] for breaking 'blackwork' (smelting residues: Robert ffille [Filley]),[59] or for cutting wood (Richard de Slymford [Slimeford, in Calstock]).[60] These could be termed 'secondary' activities which could be fulfilled by a local, unspecialised workforce. There are also recognisable tradesmen: a carpenter (John de Lyche[Leeches]),[61] a chandler (candle maker: John Lecche [Leeches]),[62] and a boat-builder (William de Clomholke/Clomhulke [Clamoak]).[63] Unfortunately payment for ore production merely appears as a total in the Exchequer

Accounts without naming the miners, although some contracts for unproductive work did give names, as when William atte Birch and 'his men' were paid for 'deadwork' (unproductive work driving adits, shafts, etc.) in 1343/44,[64] and Henry de Hullestone [Helston] was cutting and carrying wood.[65]

It is noticeable, however, that all of these people were employed in ancillary activities, and that there are no miners linked with local place-names despite forty-seven individuals being referred to in the wage rolls for the early fourteenth century.[66] This implies that the miners were not from these local places, that is they had no established family or tenurial link in contrast to people such as Richard de Slymford [of Slimeford] and John of Goueton [of Gawton] who were local farmers. These local men were, however, only employed in service-related activities supporting the mines, whereas the miners themselves were drafted in from other areas, such as Hereman de Alemannia, a German miner recorded in 1312,[67] William de Litton (1343–44),[68] Hugo de Lytton (1316),[69] Walter de Lytton (1333),[70] Walter de Warlouwe (1330s),[71] Henry Wardelow (1333–44),[72] Adam de Morneshale (1301),[73] and Hugo de Morneshale (1301–2),[74] all hailing from neighbouring settlements in the southern Peak District, and William de Dertemore (1316–25)[75] and Walter de Dertemore (1330s)[76] from Dartmoor. Significantly, while there is fifteenth- to seventeenth-century documentary evidence that tinners on south-west Dartmoor were named tenants of agricultural holdings, and often extraction was carried out on their land,[77] which supports the model of dual occupancy (part-time farmers-cum-miners), there is no such evidence for this being the case at the medieval mines of the Bere peninsula.

Accommodating the mining population

It would appear, therefore, that although some of the local farming community gained employment in supporting mining activity, they were not used in the actual extraction of the ores. So where did these miners live? Initially, the workers may have been housed in one or more temporary encampment, which, if constructed in timber, will have left few visible archaeological traces. The mining continued under the direct control of the Crown for some fifty years, however, and some more permanent solution to accommodating the imported mine workers is to be expected. It is possible that purpose-built settlements were constructed, and this may have been the case at the refining sites. When the Bere Ferrers mines were first opened in 1292 the administrative centre was located on the east side of the river Tavy at 'Martinstowe' (Maristow), where housing was built for the keeper and controller of the mines near to the silver-refining mill. This provides evidence for specifically built accommodation, but there is no mention of provision for miners or mine-labourers.[78] In 1302 the administrative centre was moved to Calstock on the Cornish side of the river Tamar (see Chapter 4), and a record of payment for moving, among other items, a small house for use by the refiners, does suggest some form of accommodation. Later in the same year, other wooden buildings, plastered and with thatched roofs, were built. These existed alongside a two-story hall, a refinery, smithy, stores, and stables, all located within and outside of the *curia* which in 1303/04 was refortified with the explicit purpose of providing security for 'men and materials'. Renovations and rebuilding to this complex is recorded until at least 1313/14 (see Chapter 4).

Back in Bere Ferrers parish, there is no evidence for any settlements along the line of the mine workings themselves that could have housed mine workers. By the beginning of the second half of the fourteenth century the keepers of the king's mines were certainly looking to place workers in properties close to the mines, and an instruction to the Sheriff of Devon dated 5 June 1360, for example, states:

> Order ...whenever required by Henry de Brusele and Richard de Colle masters of the kings's mines in that county ...to (among other things) cause houses wherein they and the workmen may be suitably lodged to be demised and delivered to them for a competent farm ... upon complaint ...that they are of times hindered ...[79]

While any of the farms and hamlets close to the mine workings could have expanded to include a few miners houses – as Henry's interpretation of page 136 in MS 3522 of the Dean and Chapter of Exeter Cathedral archives would suggest – there is no archaeological or documentary evidence for severely shrunken or indeed wholly deserted settlements. There are no field-names that suggest the sites of former habitations, and the available aerial photographic evidence does not show any evidence for shrunken or deserted settlements in the form of cropmarks or earthworks.

Another possibility is that the mine workers were housed in Bere

Alston, which is supported by either interpretation of Dean and Chapter MS 3522. Whichever way you view the list, the time allowed suggests that Bere Alston was by far the largest settlement in the parish in the late fifteenth century.[80] The size of the Bere Alston tenements is noticeably smaller than in other planned medieval towns in the South West such as West Looe, Tregony, and Newton Poppleford:[81] the width of the Bere Alston tenements appears to be 33 feet [10.1 m] (two statute perches) which is half that seen in many other planned medieval towns,[82] while the length of the tenements at Bere Alston, at 120–160 feet [37–49 m] is also far shorter than normal. At one shilling per annum in the early fourteenth century the burgage rents were a nominal acknowledgement of lordship and would have been within the means of a miner. The cost of acquiring a tenement and the status as a burgess would, however, probably be too much for an immigrant miner, particularly one who considered himself a temporary resident in the manor, when he would be expected to erect a suitable dwelling on the plot. It was, nevertheless, open to burgesses to sublet accommodation to miners and, in common with most boroughs, not all the residents of Bere Alston would be burgage tenants. Overall, it would appear that the miners employed at Bere Ferrers were mostly accommodated within the planned borough of Bere Alston, with its unusually small plots, probably making it the first specialist mining town in Britain.

Discussion

The opening of the silver mines on the Birland peninsula in the late thirteenth century, whether developing from an existing industry or not, marked a significant change in how minerals were exploited, with the emergence of direct management by the Crown. A large workforce was mustered, swelling the population of the parish considerably. Men were impressed from as far away as the Peak District and north-east Wales, though they were also drafted from tin-working areas elsewhere in Devon and Cornwall. There is good evidence to suggest that local people were employed to carry out ancillary tasks, probably supplementing their agricultural income. In part, the immigrant population may have been housed on existing farms, while the refining centres appear to have had some purpose-built accommodation. It is also possible that burgesses sublet their burgage plots within the borough at Bere Alston as miners' dwellings. Some mining families became established within the parish and took up agricultural tenancies, as is evidenced by the handful of Peak District-derived personal names appearing on the 1737 Mount Edgcumbe estate map. Comparison with the landscape of the South West as a whole suggests that the pattern of fields and settlement is, however, characteristic of the region in general and there is little to suggest that the opening of the Crown mines had any lasting impact on the landscape character: even the town at Bere Alston – probably the first mining town in Britain – fits in with a wider picture of borough foundation in the fourteenth century as lords of the manor tried to cash in on the flourishing economy. The establishment of the borough is the clearest reaction to the opening of the mines and was a seigneurial response in an effort to capitalise on the Crown enterprise. The pattern of settlements and fields that is seen today is a direct descendent of that which was largely in place by the time that the mines were opened in 1292. Both the medieval and post-medieval industries, though remarkable for the nature of their operation, had insufficient impact to produce a landscape significantly distinct from the surrounding countryside.

Chapter 7

Discussion and conclusions

An interdisciplinary approach to mining landscapes

Since the 1990s there has been a growing realisation that the whole of our landscape is of historic interest. While individual archaeological sites are legally protected through scheduling, and standing buildings of particular architectural or historical interest are listed, the now widespread practice of 'historic landscape characterisation' is designed to inform countryside managers, planners and the wider public of the time-depth that is preserved within the settlements, fields, roads, and land-uses that we use every day.[1] Across much of the country, the origins of this 'historic landscape' can be shown to date back at least as far as the Middle Ages, and there are large numbers of specialist academic groups who have an interest in researching different aspects of this period: the Society for Medieval Archaeology has one of the broadest perspectives, while more focused groups include the Castle Studies Group, the Medieval Settlement Research Group, the Society for Landscape Studies, and the Vernacular Architecture Group. This scholarly fragmentation is also seen in the study of mining, for which there is the National Association of Mining History Organisations, including groups such as the Northern Mine Research Society and the Peak District Mines Historical Society, the Association for Industrial Archaeology, the Economic History Society, the Historical Metallurgy Society, and a loose grouping of individuals in the International Mining History Congress. Each of these organisations has its own particular interest, and such specialist groups play a useful role, bringing together experts with common interests. There are, however, two fundamental problems: firstly, the lack of sufficient dialogue between these interest groups, who all too often are studying the same phenomenon – such as medieval mining – but using different source material, and secondly, that there is often a failure to see the bigger picture. Although things are changing, mining historians have all too often focused on the processes and scale of production, and the social and economic context within which it took place, while industrial archaeologists have tended to record remains on the ground, focusing particularly on the late post-medieval period, with only a minority looking at the wider landscape.

In contrast, a distinctive part of this project has been its attempt to integrate archaeological and historical approaches to the study of industry, and the structure of this book has reflected this interdisciplinary approach: instead of separate chapters dealing with the documentary material and the archaeological evidence on the ground, there have been thematic discussions of the mining process, the supply of fuel, and the wider landscape. The unusually rich documentary sources that survive from the periods of direct royal control contain not only a wealth of detail regarding many aspects of the production process itself – from development work and extraction, through to smelting and refining – but also large amounts of detail on their location that allow the mining landscape itself to be reconstructed. The mines themselves can be located, along with some of the sites where the dressed ores were smelted and refined. We know where the woodlands that supplied timber for construction and fuel were, along with some of the quays where boats moored up for the loading and unloading of goods. The lack of surviving documentary sources means we can say less about the wider agricultural landscape within which this industry was located, though a careful analysis of the historic landscape – the modern patterns of fields, roads, and settlements, that are first recorded on an estate map of 1737 – shows that in essence this too is medieval in origin: the line of mine workings clearly cuts across a pre-existing field system, as does a road linking the mines with the borough of Bere Alston founded in the early fourteenth century.

Today, the Bere Ferrers peninsula is a peaceful, rural location. Though it is not far from examples of the distinctive nineteenth-century engine houses that are so characteristic of parts of Devon and Cornwall, it is hard to imagine that the area was also once the focus of a thriving mining industry. The rich silver deposits of Bere Ferrers were mined from the late thirteenth through to the early sixteenth century, and this industry is of particular importance to the history of British mining, being the first occasion on which such operations were carried out under direct control of the Crown and in such a capital-intensive way.

The south-west of England is rich in metals which were worked in a variety of ways. Some, such as iron, were the property of the landowner, who was free to exploit the ores as he wished: while recent archaeological projects on Exmoor and the Blackdown Hills have revealed the extent to which iron was extracted during the medieval period, the industry is poorly documented. Much more is known about the tin industry because although the deposits themselves were the property of the landowner, their exploitation was subject to customary rights allowing them to be worked by any miner who paid a portion to the owner and a tax to the Crown. The better survival of documentary sources and extensive archaeological remains on and around many of the South West's granite uplands make this a relatively well-known industry, which appears to have been carried out on a part-time basis by miners-cum-farmers in what is known as dual-occupancy.

In contrast, silver – along with gold and copper – was regarded as a Crown prerogative, a right that successive kings invoked after silver was discovered in Devon during the thirteenth century. Until the late twelfth century, silver was mined in various locations in England, notably in the north Pennines but also in Mendip in Somerset. By the early thirteenth century, however, these easily accessible deposits were worked out and silver was being imported from the continent. In 1262, silver was discovered in a mine at 'la Hole', thought to be at Molland on the southern edge of Exmoor, and the Crown was quick to exercise its prerogative. Royal officials were sent, and for a few years, groups of miners, including some from central Europe, were employed in attempts to work the silver – although there is no recorded production from this mine. Prospecting appears to have continued in Devon and in 1292 new mines were opened up at Combe Martin in the north and on the manor of 'Birland' (Bere Ferrers) in the south. The former were abandoned by the Crown as unprofitable by 1296 – although there may have been some sporadic production when they were subsequently leased to various entrepreneurial interests – while those at Bere Ferrers continued to be worked directly by the Crown until 1349, and the advent of the Black Death, before they too were leased to entreprenuerial interests. What was innovative at Bere Ferrers between 1298 and 1348 was the way in which the mines came under the direct control of the Crown as a single operation allowing for the development of more capital-intensive methods. By 1298, three hundred miners were employed, many recruited from other hard-rock mining districts such as Mendip and the Peak District, with another hundred – drawn from among the local tinners – employed in drainage works. The operation of these mines was also divorced from the local customs which regulated mining elsewhere, and the produce was the Crown's to hold in full with no part passing to the land-holder.

Due to the Crown's direct management of the mines, we have unusually rich documentary sources, including financial accounts of the Exchequer. Many of the place-names they refer to can be located in the modern landscape, allowing a detailed picture of this mining landscape to be reconstructed. The mines themselves were located in a north – south oriented line between Lockridge Hill and Cleave Wood, where the ore was also broken up and sorted in order to remove the waste material. This dressed ore was then smelted to produce a lead metal rich in silver which was refined to recover the silver. Both the smelting and the refining required large amounts of fuel which meant that those activities were often carried out some distance from the mines where suitable supplies of wood and charcoal could be obtained. Overall control of the mining and processing was based at an administrative centre – or *curia* – that initially was located at or near 'Martinstowe' (Maristow) on the River Tavy. This was close to extensive areas of woodland that were the main source of fuel for the smelting and refining process, although these were soon exhausted. In 1301 the king's woods at Calstock, across the Tamar in Cornwall, were made available and the *curia* was moved there. The site at Calstock, which it is recorded as being close to the church, remained the administrative centre until at least 1316 and it had returned to 'Martinstowe' by 1320, by which time the woods there had recovered. The demand for timber – both for structural engineering within the mines and as a source for fuel – would have been considerable, and while the evidence for charcoal production is surprisingly sparse, and the over-exploitation seen at the turn of the thirteenth century was not 'good' management, the woodlands at least survived and still cloak the steep-sided valleys of both the Tamar and the Tavy valleys.

Following the Black Death the mines were included in Crown leases, but there is scant documentary evidence for production until the fifteenth century. The bullion crisis of the mid-fifteenth century stimulated renewed interest in the Bere Ferrers mines, and the 1470s saw considerable investment in innovative new pumping technology. The 16 km long Lumburn Leat built to supply water-driven pumps remains the most impressive monument to survive from the medieval silver mining industry at Bere Ferrers.

While the mines at Bere Ferrers clearly had a significant impact both in their immediate vicinity and the wider landscape, for example in the way that the Lumburn Leat cut across numerous landholdings, most places remained relatively untouched by the industry. While Avril Henry has suggested that the fifteenth-century list of tenements within Bere Ferrers parish indicates that miners were living within the numerous small hamlets that were scattered across the peninsula, Nicholas Orme has convincingly argued that this need not have been the case. The evidence, both from the list and the landscape, suggests that the majority of miners were living in the newly created borough of Bere Alston with, perhaps, a smaller number living on agricultural holdings nearer the mines. The church took a tithe on the ore mined, but the holder of the land on which the silver mines were found had no call on their produce. There was clearly the potential for conflict between the lord of the manor at Bere Ferrers and the Crown, but even though Reginald de Ferrers' petition for compensation went unheeded in the late thirteenth century, no such conflict appears to have materialised until the more unsettled years of the mid-fifteenth century. The only way that the de Ferrers could make a profit from the mines was to satisfy those demands of the new mining community that were not provided for by the Crown. The establishment of a small town with a regular market was a tested method of exploiting new opportunities provided by changed economic circumstances. The borough of Bere Alston was established in the 1290s, probably adjacent to an existing hamlet, and was superimposed on part of an established open-field system. As the service centre for the mines, Bere Alston was to grow to become the largest settlement in the area, surviving the decline in mining post-1500 to re-emerge as the focus for mining settlement in the nineteenth century. Bere Alston remains a prominent feature in the landscape, a lasting legacy of the royal silver mines, and its expansion as a modern commuter satellite for Plymouth continues to fossilise the ancient field system in which the core of the late medieval borough sits most uncomfortably. The mines to the west and south-west are today frequently linked to it by name, as the 'Bere Alston mines'. The origins are, of course, quite different. The mines were 'of Birland, or Bere Ferrers', but Bere Alston is very much a product of those mines and should perhaps be regarded as Britain's first dedicated mining town.

Notes

Notes to Chapter 1

1. 'Lake' is the local term for a stream.
2. D. Mackney et al., *Legend for the 1:250,000 Soil Map*.
3. H.G. Dines, *The Metalliferous Mining Region of South-West England*, p. 681; Leveridge et al., *Geology of the Plymouth and South-East Cornwall Area*, p. 94.
4. A further report appears in C. Smart and P. Claughton, 'The mining community and the landscape'.
5. Special interest groups of relevance to the study of the medieval landscape include the British Agricultural History Society, the Castle Studies Group, the English Place-Names Society, the Association for Industrial Archaeology, the Medieval Settlement Research Group, the Society for Landscape Studies, and the Vernacular Architecture Group.
6. COST – Cooperation in the field of Scientific and Technical research – Action A27, Landmarks: understanding pre-industrial structures in rural and mining landscapes. D. Browne and S. Hughes, *The Archaeology of the Welsh Uplands*.
7. M. Palmer, 'Industrial archaeology', pp. 1511–16.
8. D. Gwyn, *Gwynedd*, p. 19; M. Palmer, 'Industrial archaeology', p. 1518.
9. M. Nevell and J. Walker, *Lands and Lordships in Tameside* and M. Nevell, 'Industrialisation, ownership, and the Manchester methodology'. For discussion on its relevance to all aspects of industrialisation see D. Gwyn, *Gwynedd*, p. 18. The methodology would appear to be something of a blunt stick with which to pick out the origins for industrial development in a medieval mining landscape dominated by the Crown, as at Bere Ferrers, and is best suited to the free market developments in the textile industry of the eighteenth century.
10. See, for example, A.K. Hamilton Jenkin, *Mines of Devon*, and F. Booker, *Industrial Archaeology of the Tamar Valley*.
11. Mayer's study of the smelting site at Calstock is a notable exception (P. Mayer, 'Calstock and the Bere Alston silver-lead mines').

12. The Calstock Parish Archive holds transcripts and translations of documents relating to the manor and some relating to the activities of the royal mines while they were based at Calstock in the early fourteenth century. These transcripts provide an insight into the work of the mines for those without the resources to access the original documents in the National Archives (*Ibid*, p. 92).
13. Exeter, Dean and Chapter MS 3522. See A. Henry, 'Silver and salvation'; A. Henry, 'A reply' and N. Orme, 'Confession in a fifteenth-century Devon parish'.

Notes to Chapter 2

1. E. Miller and J. Hatcher, *Medieval England: Rural Society and Economic Change*, pp. 80–83.
2. P. Newman, *The Archaeology of Mining and Metallurgy*.
3. T. Greeves, 'The archaeological potential of the Devon tin industry'; G.A.M. Gerrard, *The Early Cornish Tin Industry*; S. Gerrard, 'The Dartmoor tin industry'; D. Austin *et al*., 'Tin and agriculture in the middle ages'; P. Newman, *The Archaeology of Mining and Metallurgy*, pp. 143–49.
4. F. Griffith and P. Weddell, 'Ironworking in the Blackdown Hills', pp. 27–34; Community Landscape Project reports: S. Hawken, *Recent Archaeological Discoveries at Bywood Farm*; J. Wiecken, *Ironworking in the Blackdown Hills*; C. Hawkins, *Vegetation History and Land-use Change*.
5. G. Juleff, *Early Iron-working on Exmoor*; G. Juleff, 'New radiocarbon dates for iron-working sites on Exmoor'; L.S. Bray, *The Archaeology of Iron Production*.
6. S. Paynter, *et al*., *Lead Smelting Waste*; P. Claughton *et al*., *Further work on lead/silver smelting at Combe Martin*; P. Claughton, 'Gold at North Molton'; Substrata Ltd, *Gradiometer and Resistance surveys at Mine Close*.
7. L.F. Salzman, 'Mines and stannaries'; P. Claughton, *Silver Mining*; P. Claughton, *The Combe Martin Mines*.
8. H.P.R. Finberg, 'The stannary of Tavistock'; J. Hatcher, *English Tin Production and Trade*.
9. P. Claughton, 'Mining law in England and Wales'.
10. *Cal. Close R., Hen. III*, vol. 12, p. 187.
11. *Bracton*, 3, p. 167.
12. P. Claughton, 'Mining law in England and Wales'.
13. *Pipe Roll*, PRS 1, p. 41.
14. G.R. Lewis, *The Stannaries*, pp. 233–38.
15. *Ibid*, pp. 65–84; R. Pennington, *Stannary Law*.
16. TNA: PRO, E101/261/10.

17. P. Claughton, *Silver Mining*, pp. 237–43.
18. *Ibid*, p. 230.
19. S. Stos-Gale, *Report on the Lead Isotope Analysis*.
20. A.M. Erskine, *The Accounts of the Fabric of Exeter Cathedral, 1279–1353, Part 1*, pp. 19, 50 and 78; A.M. Erskine, *The Accounts of the Fabric of Exeter Cathedral, 1279–1353, Part 2*, p. xxvii.
21. P. Claughton, 'Production and economic impact'.
22. *Pipe Roll*, PRS 30, pp. 64–65.
23. A. Raistrick and B. Jennings, *A History of Lead Mining in the Pennines*, pp. 25–26.
24. *Pipe Roll*, PRS 6, p. 6; W. Page, *The Victoria County History of Shropshire I*, p. 417.
25. N. Kirkham, *Derbyshire Lead Mining through the Centuries*, p. 100; *Cal. Liberate R.*, vol. 1, pp. 133, 394; vol. 2, p. 240.
26. A.J. Taylor, *The King's Works in Wales*, pp. 294, 349, 375, 407.
27. For examples, see C.J. Williams, 'The mining laws in Flintshire and Denbighshire'.
28. I.S.W. Blanchard, 'The miner and the agricultural community'.
29. E. Miller and J. Hatcher, *Medieval England: Towns, Commerce and Crafts*, p. 59.
30. I.S.W. Blanchard, 'Derbyshire lead production', pp. 124–25.
31. J.H. Rieuwerts, 'Lead mining in the royal forest of the Peak'; I.S.W. Blanchard, *International Lead Production and Trade*, p. 289, n. 2.
32. *Cal. Inq. Post Mortem*, vol. 8, pp. 46, 657; D. Pratt, 'Minera: township of the mines', p. 121; E.A. Lewis, 'The development of industry and commerce in Wales', p. 145.
33. J. Hatcher, *Rural Economy and Society*, p. 120.
34. I.S.W. Blanchard, 'Derbyshire lead production', p. 134.
35. *Ibid*, pp. 127–29.
36. I.S.W. Blanchard, *International Lead Production and Trade*, pp. 289–313; J.W. Gough, *Mines of Mendip*, pp. 59–63.
37. D. Pratt, 'Minera: township of the mines', pp. 122–23.
38. E.A. Lewis, 'The development of industry and commerce in Wales', p. 145.
39. C.J. Williams, 'The mining laws in Flintshire and Denbighshire', p. 62–68.
40. M.C. Gill, *The Grassington Mines*, pp. 87–124.
41. R. Burt, *The British Lead Mining Industry*, pp. 1–9.
42. For example N. Jones *et al.*, *Mountains and Orefields*, pp. 101–52.
43. See, for example, J. Barnatt and R. Penny, *The Lead Legacy*.
44. The best evidence to-date for the medieval period in the British Isles comes from Co. Waterford in Ireland where smelting slag

from Kilmacthomas has been dated to c.AD 600–700 (Neil Fairburn, 'New light on old slags – Copper smelting evidence from Irish sites' conference presentation to Historical Metallurgy Society / Mining Heritage Trust of Ireland conference on Metals and Metal-working in Ireland, Dublin, 14–16 September 2007).

45. *Letters and Papers, Hen. VIII*, vol. 16; pp. 1058, 1070, 1080, 1126 and 1132.
46. H.G. Dines, *The Metalliferous Mining Region of South-West England*; K.E. Beer and R.C. Scrivener, 'Metalliferous mineralisation', p. 125.
47. D.B. Barton, *A History of Copper Mining*, p. 9; G. Hammersley, 'Technique or economy'.
48. J.R. Harris, *The Copper King*, pp. 1–17; J. Brooke, *The Kahlmeter Journal*.
49. D. Dixon, 'Copper and gold mining in the Exmoor area'.
50. Anon, *History and Description of the New Bampfylde Copper Mine*.
51. *Cal. Fine R.* vol. 5, p. 454.
52. D. Dixon, 'Copper and gold mining in the Exmoor area', p. 43.
53. TNA: PRO, SP1/236 f. 1.
54. BGS, Dewey notebooks, 8, dated 19/6/1919; J. Rottenbury, *Geology, Mineralogy and Mining History*, p. 168.
55. Liverpool: Harold Cohen, MS 7.1, 21.
56. Anon, *History and Description of the New Bampfylde Copper Mine*.
57. D. Dixon, 'Copper and gold mining in the Exmoor area'.
58. 415BP ±35 (SUERC–7666): the calibrated dates to 2 sigma are 1426–1522, 1574–1584 and 1587–1625.
59. *Mining Journal*, vol. 23, p. 692; J. Rottenbury, *Geology, Mineralogy and Mining History*, pp. 131–32.
60. A.K. Hamilton Jenkin, *Mines of Devon*, pp. 39–40.
61. G.S. Camm, *Gold in the Counties of Devon and Cornwall*, p. 79.
62. J. Rottenbury, *Geology, Mineralogy and Mining History*, pp. 130–31.
63. P.R. Lewis and G.D.B. Jones, 'The Dolaucothi gold mines'; B. Ancel et al., *The Dolaucothi Gold Mines*.
64. TNA: PRO, E101/262/2.
65. *Cal. Close R., Ric. II*, vol. 1, 90–1.
66. T. Beare, *The Bailiff of Blackmore*, pp. 101–13.
67. BL Add. MS 24513, f95.
68. J. Calvert, *The Gold Rocks of Great Britain and Ireland*, p. 307.
69. P. Claughton, 'Gold at North Molton'.
70. e.g. G.R. Lewis, *The Stannaries*; H.P.R. Finberg, 'The stannary of Tavistock'; J. Hatcher, *English Tin Production and Trade*; T. Greeves,

'The archaeological potential of the Devon tin industry'; S. Gerrard, 'Streamworking in medieval Cornwall'; D. Austin *et al.*, 'Tin and agriculture in the middle ages'; J.R. Maddicott 'Trade, industry and the wealth of King Alfred'; S. Gerrard, 'The Dartmoor tin industry'; T. Greeves and P. Newman, 'Tin-working and land-use in the Walkham Valley'; P. Newman, 'Tinners and tenants on south-west Dartmoor'; S. Gerrard, 'The early south-western tin industry'; S. Gerrard, *The Early British Tin Industry*; P. Newman, 'Tin-working and the landscape of medieval Devon'.

71. See, for example, P. Newman, *The Archaeology of Mining and Metallurgy*.
72. V.R. Thorndycraft *et al.*, 'Tracing the record of early alluvial tin mining on Dartmoor'; V.R. Thorndycraft *et al.*, 'An environmental approach to the archaeology of tin mining on Dartmoor'; V.R. Thorndycraft *et al.*, 'Alluvial records of medieval and prehistoric tin mining on Dartmoor'.
73. K.E. Beer and R.C. Scrivener 'Metalliferous mineralisation'.
74. J. Hatcher, *English Tin Production and Trade*, pp. 49–88.
75. J. Hatcher, 'Myths, miners and agricultural communities'.
76. H.P.R. Finberg, 'The stannary of Tavistock', pp. 164–65; R. Pennington, *Stannary Law*, pp. 80–81.
77. V.R. Thorndycraft *et al.*, 'Tracing the record of early alluvial tin mining on Dartmoor', p. 92.
78. G.R. Lewis, *The Stannaries*, p. 163.
79. A. Knighton, 'River adjustment to changes in sediment load'.
80. T. Greeves, 'The archaeological potential of the Devon tin industry'.
81. S. Gerrard, 'Streamworking in medieval Cornwall', pp. 7–31; S. Gerrard, 'The Dartmoor tin industry', pp. 175–80; S. Gerrard, 'The early south-western tin industry', pp. 70–76; P. Newman 'The Moorland Meavy'.
82. S. Gerrard, *The Early British Tin Industry*, pp. 79–80.
83. V.R. Thorndycraft *et al.*, 'Tracing the record of early alluvial tin mining on Dartmoor'; V.R. Thorndycraft *et al.*, 'An environmental approach to the archaeology of tin mining on Dartmoor'; V.R. Thorndycraft *et al.*, 'Alluvial records of medieval and prehistoric tin mining on Dartmoor'; for an earlier application on Mendip, Somerset, see M. Macklin, 'Floodplain sedimentation in the upper Axe valley'.
84. V.R. Thorndycraft *et al.*, 'Tracing the record of early alluvial tin mining on Dartmoor', p. 98.
85. V.R. Thorndycraft *et al.*, 'An environmental approach to the archaeology of tin mining on Dartmoor', pp. 21–23; V.R. Thorndycraft

et al., 'Alluvial records of medieval and prehistoric tin mining on Dartmoor', pp. 232–33.

86. V.R. Thorndycraft *et al.*, 'Tracing the record of early alluvial tin mining on Dartmoor', pp. 98–100; V.R. Thorndycraft *et al.*, 'Alluvial records of medieval and prehistoric tin mining on Dartmoor', p. 233; T. Greeves 'Four Devon stannaries', p. 63, table 1.
87. S. Gerrard, 'The Dartmoor tin industry', p. 175; S. Gerrard, 'The early south-western tin industry', p. 69.
88. S. Gerrard, *The Early British Tin Industry*, p. 29.
89. P. Newman, 'Tin-working and the landscape of medieval Devon', pp. 133–35.
90. T. Greeves, 'The archaeological potential of the Devon tin industry', p. 89.
91. R.F. Tylecote, *The Prehistory of Metallurgy in the British Isles*, pp. 44–47.
92. G.R. Lewis, *The Stannaries*, p. 134.
93. B. Earl, 'Melting tin in the West of England'; S. Gerrard, *The Early British Tin Industry*, pp. 129–39.
94. A. Malham *et al.*, 'Tin smelting slags from Crift Farm'.
95. T. Greeves 'The archaeological potential of the Devon tin industry'; T. Greeves, 'Four Devon Stannaries'.
96. P. Newman 'Tinners and tenants on south-west Dartmoor'.
97. H.P.R. Finberg, 'The stannary of Tavistock', pp. 173–82.
98. S. Gerrard, *The Early British Tin Industry*, pp. 50–53.
99. J. Hatcher, 'Myths, miners and agricultural communities'; I.S.W. Blanchard, 'The miner and the agricultural community'; I.S.W. Blanchard, 'Rejoinder; Stannator Fabulosus,'.
100. J. Hatcher, 'Myths, miners and agricultural communities', pp. 56–57; S. Gerrard, *The Early British Tin Industry*, p. 53.
101. S. Gerrard, *The Early British Tin Industry*, p. 50.
102. H.S.A. Fox, 'Medieval Dartmoor'.
103. B. Gearey *et al.*, 'The landscape context of medieval settlement'.
104. V.R. Thorndycraft *et al.*, 'An environmental approach to the archaeology of tin mining on Dartmoor', pp. 23–24; D. Austin *et al.*, 'Farms and fields in Okehampton Park'; D. Austin and M. Walker, 'A new landscape context for Houndtor'.
105. C.A. Roberts and D.D. Gilbertson, 'The vegetational history of Pile Copse'; V.R. Thorndycraft *et al.*, 'An environmental approach to the archaeology of tin mining on Dartmoor'.
106. T. Greeves 'The archaeological potential of the Devon tin industry', pp. 85–86.
107. H.R. Schubert, *History of the British Iron and Steel Industry*, pp. 81–87.

108. E. Miller and J. Hatcher, *Medieval England: Towns, Commerce and Crafts*, pp. 61–63; S. Moorhouse, 'Iron production', p. 779.
109. G. McDonnell, 'Monks and miners'; G. Astill, *A Medieval Industrial Complex*, pp. 300–2.
110. H.R. Schubert, *History of the British Iron and Steel Industry*, pp. 94–98.
111. S. Pollard and D.W. Crossley, *The Wealth of Britain*, p. 44.
112. H.R. Schubert, *History of the British Iron and Steel Industry*, pp. 111–15, 145; E. Miller and J. Hatcher, *Medieval England: Towns, Commerce and Crafts*, p. 63; D. Crossley 'Medieval iron smelting'.
113. M. Zell, *Industry in the Countryside*, pp. 237–38.
114. R.F. Tylecote, *The Prehistory of Metallurgy in the British Isles*, p. 125.
115. R.F. Tylecote, 'Iron Smelting in Pre-Industrial Communities', pp. 116–17; S. Moorhouse, 'Iron production'.
116. K.C. Dunham, *Geology of the Northern Pennine Orefield*, p. 4.
117. H. Cleere and D. Crossley, *The Iron Industry of the Weald*, pp. 15–21, 98–99.
118. F. Griffith and P. Weddell, 'Ironworking in the Blackdown Hills'.
119. *Domesday Book, Somerset*, 1,4; 3,1; 17,3; 19,4; 19,27; 19,65; 21,75–76; *Domesday Book, Index of Subjects*, p. 98.
120. G. Juleff, *Early Iron-working on Exmoor*, p. 13.
121. N. Nayling, *The Magor Pill Medieval Wreck*, pp. 105–15; J.R.L. Allen, 'A possible medieval trade'; J.R.L. Allen, 'A medieval pottery assemblage'.
122. J.R.L. Allen, 'A possible medieval trade'.
123. S.A. Moorhouse 'Iron production', p. 755.
124. S.J. Reed, *Archaeological Recording of an Iron Ore Extraction Pit*.
125. F. Griffith and P. Weddell, 'Ironworking in the Blackdown Hills'; S.J. Reed, *Archaeological Recording of an Iron Ore Extraction Pit*.
126. S. Hawken, *Recent Archaeological Discoveries at Bywood Farm*; C. Hawkins, *Vegetation History and Land-use Change*.
127. see G. Juleff *Early Iron-working on Exmoor*; H. Riley and R. Wilson-North, *Field Archaeology of Exmoor*, pp. 111–13; L.S Bray, *The Archaeology of Iron Production*.
128. R. Burt *et al.*, *The Devon and Somerset Mines*.
129. *Domesday Book, Devon*, 1,27.
130. T. Westcote, *View of Devonshire*.
131. Bodleian MS Top Devon 62.
132. *Mining World*, V 1873, p. 854.
133. J.A. Cannell, *The Archaeology of Woodland Exploitation*, p. 121.
134. N. Berry, *Horner Wood*, p. 9 et seq; J.A. Cannell, *The Archaeology of Woodland Exploitation*, p. 121.
135. Lee Bray pers. comm.

136. *Cal. Pat R., Edw. VI*, 3, pp. 344–45.
137. NDA, 'Bristol Channel Shipping'.
138. Bodleian MS Top Devon 62.
139. H. Riley and R. Wilson-North, *Field Archaeology of Exmoor*, pp. 112–13.
140. J.A. Cannell, *The Archaeology of Woodland Exploitation*, p. 77.
141. R. McDonnell, *Hawkridge Ridge Wood*; R. McDonnell, *Preliminary Archaeological Assessment of Eleven Woodlands*; J.A. Cannell, *The Archaeology of Woodland Exploitation*, p. 182.
142. J.A. Cannell, *The Archaeology of Woodland Exploitation*, p. 182.
143. L.S. Bray, *The Archaeology of Iron Production*.
144. H. Summerson, *Crown Pleas*, p. 21.
145. F. Griffith and P. Weddell, 'Ironworking in the Blackdown Hills', p. 33.
146. C. Hawken, *Recent Archaeological Discoveries at Bywood Farm*, p. 21; F. Griffith and P. Weddell, 'Ironworking in the Blackdown Hills', pp. 31–32; C. Hawkins, *Vegetation History and Land-use Change*, p. 12.
147. C. Hawkins, *Vegetation History and Land-use Change*, p. 12.
148. T. Mighall and F. Chambers, 'Early ironworking and its impact on the environment'.
149. G. Juleff 'The Dark Ages remain a mystery'.

Notes to Chapter 3

1. T.M. Hall, 'On the mineral localities of Devonshire', p. 340.
2. P.T. Craddock, *Early Metal Mining and Production*, pp. 221–28.
3. H.W.W. Ashworth, *Report on Romano-British settlement and metallurgical site*; D. Elkington, 'The Mendip lead industry'; and R.G.J. Williams, 'St Cuthbert's Roman mining settlement'.
4. J.C. Edmunson, 'Mining in the later Roman empire and beyond'.
5. M. Todd, *Roman Mining on Mendip*, pp. 75–82.
6. *Ibid*, pp. 77–78; J. Bayley and K. Eckstein, *Metal-working Debris from Pentrehyling Fort*.
7. P. Spufford, *Money*, p. 9.
8. D.M. Metcalf, 'The ranking of boroughs', p. 205.
9. P. Spufford, *Money*, maps 10A and 10B.
10. *Ibid*, p. 94.
11. M. Blackburn, 'Coinage and currency under Henry I'.
12. Sir J. Craig, *The Mint*, App. 1.
13. P. Spufford, *Money*, p. 209. See also *ibid.*, p. 295, Table 5, for the long-term changes in twelve major currencies 1300–1500.
14. See, for example, M. Ponting, et al. 'Fingerprinting of Roman mints using laser-ablation'.
15. M. Allen, 'Henry II and the English Coinage'.

16. M. Allen, 'The volume of the English currency'.
17. P. Spufford, *Money*, Map 10A; *Domesday Book, Derbyshire* 1,15.
18. *Magnus Rotulus Scaccarii de anno 31 Henrici I*, p. 142.
19. *Chronicles of the Reigns of Stephen, Henry II and Richard I*, vol. 4, p. 123.
20. G.W.S. Barrow, *The Acts of Malcolm IV*, pp. 111–12 and nos 39–40.
21. See J. Wilson, *County History of Cumberland*, for translations.
22. P. Claughton, 'Production and economic impact'.
23. C. Tolan-Smith, *Landscape Archaeology in Tynedale*; A. Winchester, *The Harvest of the Hills*.
24. *Calendar of Documents relating to Scotland*, vol. 1, p. 125.
25. *Pipe Rolls*, PRS NS 2, p. 55.
26. K.C. Dunham, *et al.*, 'Rich silver-bearing lead ores in the northern Pennines?'.
27. *Calendar of Documents relating to Scotland*, vol. 3, p. 295; *Cal. Inq. Misc.*, vol. 3, p. 81.
28. See discussion in I.S.W. Blanchard, 'The miner and the agricultural community'; and J. Hatcher, 'Myths miners and the agricultural community', along with I.S.W. Blanchard's 'Rejoinder, Stannator Fabulosus'; and I.S.W Blanchard, 'Labour productivity'.
29. *Pipe Rolls*, PRS NS 30, p. 154.
30. P. Claughton, 'Production and economic impact', p. 148.
31. R.C. Hoare, *Giraldus Cambrensis*, p. 129.
32. W.C. Wells, 'The Shrewsbury Mint' and J.D. Brand, 'Some short cross questions'.
33. TNA: PRO, SC1/48/177. J.W. Gough, *Mines of Mendip*, pp. 51–53.
34. F. Téreygeol, 'Les mines d'argent carolingiennes de Melle'.
35. P. Spufford, *Money*, pp. 74–105 and Map 9.
36. J.U. Nef, 'Mining and metallurgy in medieval civilisation', p. 435.
37. P. Spufford, *Money*, pp. 109–31.
38. S. Cirkovic, 'The production of gold, silver and copper', p. 52.
39. I.S.W. Blanchard, *International Lead Production and Trade*, pp. 15–22.
40. C.M. Cipolla, *Before the Industrial Revolution*, p. 175.
41. M. Allen, 'The volume of the English currency', pp. 601–3.
42. *Cal. Close R., Hen. III*, vol. 12, p. 187.
43. *Ibid.*, vol. 12, pp. 214 and 227–28; *Cal. Pat R., Hen. III*, vol. 5, pp. 249, 255, and 256; *Cal. Liberate R.*, vol. 5, p. 120; *ibid.*, vol. 6, item 2325.
44. *Cal. Pat R., Hen. III*, vol. 5, p. 304; *Cal. Close R., Hen. III*, vol. 7, pp. 349 and 406–7; *Cal. Liberate R.*, vol. 5, p. 246; *ibid*, vol. 6, item 1252.
45. See for example *Cal. Pat R., Edw. I*, vol. 1, p. 161 and vol. 2, p. 322.

46. TNA: PRO, E101/260/4, 5 and 6.
47. R.A. Brown, et al., *The History of the King's Works, Vol. I: The Middle Ages*, pp. 54 and 180–82.
48. TNA: PRO, E101/260/17.
49. BL, Add. MS 24770, f220.
50. C.G. Crump and C. Johnson, 'Tables of bullion', p. 9.
51. R.W. Kaeuper, 'The Frescobaldi of Florence and the English Crown', p. 60.
52. *Cal. Pat R., Edw. I*, vol. 4, p. 513. The term 'last' is an undefined measure; however, the later use of the term *ladam* suggests that the measure was the 'load', i.e. 9 dishes, of ore (*Cal. Close R. Edw. II*, vol. 1, p. 95) – see glossary.
53. R.W. Kaeuper, 'The Frescobaldi of Florence and the English Crown', p. 62.
54. P. Spufford, *Money*, p. 127, quoting 'mint accounts'.
55. TNA: PRO, E101/260/19 f2.
56. TNA: PRO, E101/261/22.
57. TNA: PRO, E101/263/7.
58. TNA: PRO, E159/91 f110; *Cal. Memo. R.*, 1326–1327, items 1211, 1858, 1920 and 1925.
59. TNA: PRO, C143/192/10.
60. P. Claughton, *The Combe Martin Mines*.
61. *Cal. Close R. Edw. III*, vol. 10, pp. 663–64; TNA: PRO, E101/263/12.
62. *Cal. Close R. Edw. III*, vol. 11, p. 37.
63. TNA: PRO, SP1/236.
64. P. Claughton, *The Combe Martin Mines*, pp. 9–11.
65. Ibid., pp. 11–12, Paynter et al., *Lead Smelting Waste*.
66. *Cal. Fine R.*, vol. 6, p. 117.
67. BL, Add. MS 24513 f95 and f96; TNA: PRO, E101/265/10 and 11, *Cal. Pat R., Hen. VI*, vol. 5, p. 533.
68. *Cal. Pat R., Hen. VI*, vol. 5, p. 215.
69. I.S.W. Blanchard, *International Lead Production and Trade*, p. 302.
70. TNA: PRO, E101/263/7.
71. *Cal. Pat R., Hen. VI*, vol. 5, pp. 467–68.
72. *Cal. Pat R., Hen. VI*, vol. 6, pp. 158 and 291; *Cal. Pat R., Edw. IV*, p. 19.
73. Devon Record Office 1815Z/PZ 1–3.
74. E. Stuart, *Lost Landscapes of Plymouth*, pp. 29 and 95; Cornwall Record Office ME2424.
75. C. Bartels, et al. *Kupfer, Blei und Silber aus dem Goslarer Rammelsberg*; P. Claughton, *The Combe Martin Mines*; G. Boon, *Cardiganshire Silver*.

76. E. Plowden, *The commentaries*, vol. 1, pp. 310–40.
77. *The Statutes*, vol. 1, pp. 425 and 434–36.
78. D. Bick, *Waller's Description of the Mines in Cardiganshire*, pp. 7–8.
79. R. Burt, *The British Lead Mining Industry*, p. 7.

Notes to Chapter 4

1. See, for example, the autonomy allowed to the coal miner at the height of capital investment in the early nineteenth century (J.A. Jaffe, *Market Power*, p. 48).
2. TNA: PRO, E101/263/7, entry for 2 August.
3. Virtually no work has been done on the archaeology of underground lead/silver workings which can be clearly dated to the medieval period. For useful comparisons it is necessary to look to the European continent, in particular France – see, for example, M-C. Bailly-Maître, *L'argent, du Minerai au Pouvoir dans la France Médiévale*.
4. See, for example, the investigations carried out by the Plymouth Caving Group, for Devon County Council, on workings under the road at Gullytown.
5. TNA: PRO, E101/260/4.
6. Pete Budge (Bere Alston), pers. comm. 5 April 2007.
7. National Library of Wales, *G.E. Owen Papers*, Box 54.
8. *Cal. Pat R., Hen. VI*, vol. 5, p. 571.
9. TNA: PRO, E101/261/10.
10. P.T. Craddock, *Early Metal Mining and Production*, pp. 33–37.
11. L.F. Salzman, 'Mines and stannaries', p. 73.
12. TNA: PRO, SC1/48/61.
13. The Frescobaldi accounted for the issues of the mines in Devon from 18 April, 27 Edw. I to 26 February, 29 Edw. I (*Cal. Pat R., Edw. II*, vol. 1, pp. 234–35).
14. TNA: PRO, E101/261/12.
15. TNA: PRO, E101/260/30 and 261/12.
16. TNA: PRO, E101/262/13. The use of crosscutting drainage adits was not new, they had been used as early as the second (La Tene) Iron Age in the gold mines of the Limousin region of France (B. Cauuet, *Les Mines d'Or*, pp. 22–23, and 'Celtic gold mines', pp. 231–32.
17. Based on drivage rates in the sixteenth century quoted in J.A. Buckley, *The significance and role of adits*, p. 27.
18. DRO, 4672A/HS/79E.
19. TNA: PRO, E101/260/19.
20. P. Claughton, 'Silver-lead – a restricted resource'.
21. TNA: PRO, E101/263/7. The 'grove' element in the place-name being the term used for a mine working in the north of England.

22. TNA: PRO, E101/263/8 and E101/263/10. For details of the earlier trial see L.F. Salzman, 'Mines and stannaries', pp. 84–85.
23. TNA: PRO, E101/263/8. *Cal. Pat R., Edw. I*, vol. 3, p. 577. *Cal. Close R., Edw. III*, vol. 7, p. 406. In 1498 there was also a futile trial for silver at Brushford, in Devon, in an area with little evidence for significant mineralisation (*Cal. Pat R., Hen. VII*, vol. 2, p. 135).
24. *Cal. Fine R.*, vol. 2, p. 126. *Cal. Close R., Edw. II*, vol. 2, p. 52.
25. *Cal. Pat R., Hen VII*, vol. 2, p. 135.
26. TNA: PRO, E101/260/27, particularly the entry for 24 October; 260/30 and 261/21.
27. TNA: PRO, E101/261/12.
28. J. Langdon and J. Masschaele, 'Commercial activity and population growth in medieval England', pp. 69–77.
29. L.F. Salzman, 'Mines and stannaries', p. 73.
30. Calstock Parish Archive, transcripts of Duchy of Cornwall 472 and TNA: PRO, *Augmentation Office* E306/2/1. The name would probably have originated from Smalley near Ripley.
31. DRO, 4672A/HS/79E; *Mining Journal*, vol. 16, p. 415.
32. DRO, tithe award for Tamerton Foliot parish; TNA: PRO, E101/265/18. In the case of Higesbale, the 'bale' element is a common place-name element indicating a smelting site in Yorkshire and the north Pennines, and might be an imported name: equally it could be a corruption of 'ball', a round hill.
33. TNA: PRO, E101/260/22, see, for example, the entries for 2 February, 20 July and 3 August, 30 Edw. I.
34. TNA: PRO, E372/145.
35. TNA: PRO, E101/260/22, entry for 31 August, 31 Edw. I.
36. TNA: PRO, E159/92 f96 (transcript in the Calstock Parish Archive).
37. See, for example, TNA: PRO, E101/263/5, entry for 8 February and E101/260/22, entry for 26 October.
38. TNA: PRO, E101/260/30.
39. TNA: PRO, E101/261/22.
40. Manually operated bellows continued to be used throughout the first half of the fourteenth century and the *slaggemill* referred to in the 1330s (L.F. Salzman, 'Mines and Stannaries', p. 76) is probably the device used for crushing smelting residues or slag.
41. TNA: PRO, E101/260/19 f1.
42. TNA: PRO, E101/260/22 and E101/260/7, particularly entry for 3 October 1304.
43. TNA: PRO, E101/260/19 f2.
44. TNA: PRO, E101/260/30, entry for 21 May.
45. TNA: PRO, E101/260/22, entry for 5 October.

46. TNA: PRO, E 101/260/22, 31 Edw. I, 10, 17, 24 August and 7 September etc. 1303.
47. TNA: PRO, E101/260/26 Wage Roll 1304–5 entry for 18 July.
48. TNA: PRO, E101/260/30 Wages Roll 1306, entry for 23 July.
49. For example TNA: PRO, E101/260/22, entry for 14 September 1302.
50. *Ibid*, entry for 3 August 1302.
51. *Ibid*.
52. TNA: PRO, E101/260/19 f4.
53. TNA: PRO, E101/260/27, entries for 29 August, 12 September and 24 October.
54. TNA: PRO, E101/260/19 f5.
55. TNA: PRO, E101/260/26, entries for 13 June and 4 July.
56. TNA: PRO, E101/261/10.
57. TNA: PRO, E372/161 f52.
58. See G. Agricola, *De Re Metallica*, p. 467.
59. P.T. Craddock, *Early Metal Mining and Production*, pp. 221–33.
60. TNA: PRO, E101/260/4. See also Sir E. Smirke, 'Supplementary notes relating to the working of the silver mines in Devon', p. 319.
61. TNA: PRO, E101/262/4.
62. P. Claughton, *Silver Mining*, p. 173.
63. J. Bayley and K. Eckstein, *Metal-working Debris from Pentrehyling Fort*.
64. TNA: PRO, E101/260/26, entry for 8 August.

Notes to Chapter 5

1. J.R. Hunter, 'The medieval glass industry', p. 149; S.A. Moorhouse, 'The medieval pottery industry', p. 119; J. Cherry, 'Pottery and tile', p. 189; J. Cherry, 'Leather', p. 301; R.D. Berryman, *Use of the Woodlands*, pp. 10–37.
2. P. Claughton, *Silver Mining*, pp. 210–11; M.C. Gill, *The Grassington Mines*, p. 24.
3. P. Claughton, *Silver Mining*, p. 177; for example, see *Cal. Pat. R. Edw. III*, vol. 4, p. 286.
4. TNA: PRO, E101/261/15; E101/261/21.
5. M. Jones, 'South Yorkshire's ancient woodland', p. 33. Ordinance for the Preservation of Woods Act, 1543.
6. O. Rackham, *Trees and Woodland*, p. 63; B.M.S. Campbell and K. Bartley, *England on the Eve of the Black Death*, Map 9.20.
7. J. Wheeler, *The Implications of Iron-Working*.
8. D. Crossley, 'White coal and charcoal'.
9. J. Birrell, 'The forest and the chase', p. 36; D.L. Farmer, 'Woodland and pasture sales', pp. 111–12.
10. E. Searle, *Lordship and Community*, p. 266.

11. TNA: PRO, E101/261/12.
12. The purchase of refinery ash was recorded in 1294 (TNA: PRO, E101/260/4), 1306 (TNA: PRO, E101/260/30), 1308 (TNA: PRO, E101/261/10), 1310 (TNA: PRO, E101/261/12), 1313 (TNA: PRO, E101/261/15 and 261/15 f2), 1343 (TNA: PRO, E101/263/5), 1344 (TNA: PRO, E101/263/8), 1349 (TNA: PRO, E101/263/11) and 1481 (TNA: PRO, E101/266/25).
13. J.A. Cannell, *The Archaeology of Woodland Exploitation*, p. 24.
14. R.W. Kaeuper, 'The Frescobaldi of Florence and the English Crown', p. 61.
15. B.M.S. Campbell and K. Bartley, *England on the Eve of the Black Death*, p, 157.
16. J. Bond, 'Medieval charcoal-burning', p. 292.
17. J. Wheeler, *The Implications of Iron-Working*, p. 377.
18. J.A. Cannell, *The Archaeology of Woodland Exploitation*, p. 29.
19. O. Rackham, *The History of the Countryside*, p. 91; O. Rackham, *Trees and Woodland*, p. 85.
20. J. Cherry, 'Pottery and tile', p. 189; S.A. Moorhouse, 'The medieval pottery industry', p. 119.
21. *Cal. Close R., Edw. I*, vol. 4, p. 433.
22. TNA: PRO, E368/85 m.204.
23. BL, Add. 24770, f267.
24. *Cal. Inq. Misc.*, vol. 2, item 1650.
25. Half of the 300 acres were used in a single year during the Frescobaldi's lease; TNA: PRO, E368/73 mm. 17d, 18.
26. J. Bond, 'Medieval charcoal-burning', 279.
27. J. Bond, 'Medieval charcoal-burning', 279.
28. J. Bond, 'Medieval charcoal-burning', p. 279–80.
29. H. Summerson, *Crown Pleas*, p. 21.
30. B. Groenewoudt, 'Charcoal burning and landscape dynamics'; limited evidence for the use of pits '*fossae*' in the Forest of Dean in the thirteenth and fourteenth century (C. Hart, *Industrial History of Dean*, p. 324); J. Bond, 'Medieval charcoal-burning', p. 286.
31. DCC HER record numbers 67207, 67213, 67214, 67215 and 67216.
32. C. Buck, *Devon Great Consols*, 236 and figure 38.
33. J. Gover et al., *The Place-Names of Devon*, p. 223; J. Field, *English Field-Names*, p. 26; J. Field, *A History of English Field-Names*, p. 61; J. Bond, 'Medieval charcoal-burning', p. 290.
34. J. Bond, 'Medieval charcoal-burning', p. 290.
35. One quarter comprised eight bushels and each bushel eight gallons dry volume.
36. TNA: PRO, E101/263/7 19 July 1343.
37. TNA: PRO, E101/263/10 7 February 1344.

38. TNA: PRO, E101/263/5; E101/263/8.
39. TNA: PRO, E101/263/11.
40. *Ibid.*
41. TNA: PRO, E101/260/5.
42. See, for example, TNA: PRO, E101/261/10.
43. J. Hatcher, *Towards the Age of Coal*, pp. 418–25.
44. 'For the wages of one woman sieving charcoal at the smithy' in 1343 (TNA: PRO, E101/263/5).
45. TNA: PRO, E101/266/25.
46. In 1306 a barge was hired to move 11 quarters of sea coal from Sutton to Calstock at a cost of 2s. (TNA: PRO, E101/260/30 entry for 20 August).
47. J.A. Cannell, *The Archaeology of Woodland Exploitation*, p. 29.
48. TNA: PRO, E101/260/30.
49. J. Bond, 'Medieval charcoal-burning', p. 277.
50. J. Birrell, 'The peasant craftsman'.
51. TNA: PRO, E101/261/10.
52. TNA: PRO, E101/261/15.
53. TNA: PRO, E101/266/25.
54. A. Jones, 'Flintshire ministers' accounts', appendix A.
55. TNA: PRO, SC1/48/61 – for a transcript see *Recuiel de lettres Anglo-Francaises*.
56. WRO 366/1.
57. See G. Agricola, *De Re Metallica*, pp. 177 and 185.
58. TNA: PRO, E101/266/25.
59. *Ibid.*
60. G. Hollister-Short, 'On the origins of the suction-lift pump'.
61. A.R. Myers, *The Household of Edward IV*, p. 282.
62. *Cal. Pat R., Edw. IV – Edw. V – Ric. III*, p. 213.
63. F. Booker, *The Industrial Archaeology of the Tamar Valley*.
64. A.K. Hamilton Jenkin, *Mines of Devon*.
65. P. Claughton, 'Lumburn Leat'; P. Claughton, *Silver Mining*.
66. TNA: PRO, E101/266/25.
67. WYAS (Leeds) DB 178/60.
68. There is some indication that furnace smelting was being carried out at Maristow in the early fourteenth century. In December 1302 ore and charcoal were shipped to St Martins (TNA: PRO, E101/260/22, 22 December, 30 Edw. I).
69. TNA: PRO, E101/260/22, 9 November, 31 Edw. I.
70. Wages of 4d. per day were paid for the removal of the turnbole from Buckland in 1302 (TNA: PRO, E101/260/22).

Notes to Chapter 6

1. P. Claughton, *Silver Mining*, appendix 4.
2. S. Rippon, *Historic Landscape Analysis*; S. Rippon, *Beyond the Medieval Village*.
3. Discussion of 'Historic Landscape Characterisation' can be found in S. Rippon, *Historic Landscape Analysis*, and the special collection of papers in volume 8 (part 2) of the journal *Landscapes* (edited by David Austin, Stephen Rippon and Paul Stamper).
4. S. Rippon, 'Landscapes of pre-medieval occupation', S. Rippon, 'Emerging regional variation', S. Rippon, *Beyond the Medieval Village*; S.J. Rippon et al. 'Beyond villages and open fields'.
5. For the project the second Edition maps, surveyed in 1839–88 and revised in 1903, were used.
6. S. Rippon, *Historic Landscape Analysis*, fig. 26; P. Herring, 'Cornish strip fields'.
7. 1845 Tithe map DRO reference.
8. A. Henry, 'Silver and salvation'; CRO: ME2423.
9. For example see M.W. Beresford, 'Dispersed and grouped settlement'; A.H. Shorter et al., *Southwest England*, p. 105; T. Rowley, *Villages in the Landscape*, p. 5; B.K. Roberts and S. Wrathmell, *An Atlas of Rural Settlement*, p. 57; S.J. Rippon et al., 'Beyond villages and open fields'.
10. S. Rippon, 'Emerging regional variation'; S. Rippon, *Beyond the Medieval Village*; S.J. Rippon et al., 'Beyond villages and open fields'.
11. E. Stuart, *Lost Landscapes of Plymouth*, colour plate 2, and p. 77 (original in the Marquis of Salisbury's collection at Hatfield House).
12. CRO: ME2423.
13. J. Field, *History of English Field-Names*, p. 25.
14. L. Toulmin Smith (ed.), *The Itinerary of John Leland*, p. 212.
15. *Domesday Book*, Devon 15, 46.
16. J.E.B. Gover et al., *The Place-Names of Devon*, p. 223 state that 'It is impossible to offer any satisfactory explanation of this name. It is clearly not the common Devon *beere, beare*, from OE **bearu** ... neither is it NCy [north country] **byre**, 'byre'. E. Ekwall, *The Concise Oxford Dictionary of English Place-Names*, suggests that it is derived from 'OE *byrh*, a dat. form of *burh*. The *-h* would be particularly apt to be lost in the compound Byrhland, whence Birland'.
17. R. Coates et al., *Celtic Voices*, p. 289.
18. D. Lysons and S. Lysons, *Magna Britannia*, pp. 27–47.
19. W.G. Hoskins, *Devon*, p. 276; R. Higham, *The Castles of Medieval Devon*, p. 52.
20. R. Higham, 'Castles in Devon', p. 72.

21. W.G. Hoskins, *Devon*, p. 276.
22. A. Henry, 'Silver and salvation'.
23. A name appearing on nineteenth-century Ordnance Survey First Edition Six Inch maps.
24. S. Rippon, *Landscape, Community and Colonisation*, p. 189 n. 10.
25. M. Beresford and H.P.R. Finberg, *English Medieval Boroughs*, p. 87.
26. P. Claughton, *Silver Mining*, pp. 235–36.
27. *Cal. Charter R.*, vol. 2, p. 463.
28. BL Harl MS 6126, f68; M.W. Beresford and H.P.R. Finberg, *English Medieval Boroughs*, p. 89.
29. For example, *Cal. Pat R., Edw. II*, vol. 1, p. 14. See also P. Claughton, *Silver Mining*, pp. 232–33.
30. A. Everitt, 'The marketing of agricultural produce', p. 471.
31. M.W. Beresford and H.P.R. Finberg, *English Medieval Boroughs*, table 1, pp. 41–43.
32. M.W. Beresford and H.P.R. Finberg, *English Medieval Boroughs*, p. 98; F. Griffith, *Devon's Past*, p. 83.
33. M.W. Beresford and H.P.R. Finberg, *English Medieval Boroughs*, table 7, table 8.
34. M.W. Beresford and H.P.R. Finberg, *English Medieval Boroughs*, pp. 51–54.
35. S. Rippon, 'Landscapes of pre-medieval occupation'; S.J. Rippon et al., 'Beyond villages and open fields'.
36. H.S.A. Fox, 'Medieval urban development', p. 402.
37. J. Field, *History of Field-Names*, p. 11.
38. A.J.C. Beddow, *A History of Bere Ferrers Parish*, p. 16.
39. TNA: PRO, E101/260/30.
40. H.P.R. Finberg, *Tavistock Abbey*, pp. 50–52.
41. P. Herring, 'Cornish strip fields', p. 59.
42. A. Rowe, *Buckland Monochorum*; T. Greeves and A. Rowe, 'Four deserted medieval settlements'.
43. T. Greeves and A. Rowe, 'Four deserted medieval settlements', p. 72; G. Oliver, *Monasticon Diocesis Exoniensis*.
44. J.E.B. Gover et al., *The Place-Names of Devon*, p. 224.
45. A.J.C. Beddow, *A History of Bere Ferrers Parish*, p. 27.
46. For example H.S.A. Fox, 'Occupation of the land', p. 164; C.G. Henderson and P.J. Weddell, 'Medieval settlements on Dartmoor', p. 131; S.J. Rippon et al., 'Beyond villages and open fields', p. 63.
47. P. Herring, 'Cornish strip fields'; S. Turner, *Ancient Country*, pp. 48–56.
48. TNA: PRO, SC8/322/E550 dated 1295–1320.
49. TNA: PRO, SC8/111/5544.
50. H.P.R. Finberg, 'Morwell', p. 164.

51. H.P.R. Finberg, 'The open field in Devon', p. 148.
52. A. Henry, 'Silver and Salvation'.
53. N. Orme, 'Confession in a fifteenth-century Devon parish'.
54. A. Henry, 'A reply'.
55. For example see S. Rippon, *Landscape, Community and Colonisation*.
56. TNA: PRO, E101/260/22, 31 Edw. I, 13 July 1303.
57. TNA: PRO, E101/261/25, 16–18 Edw. II, 4 May 1323.
58. TNA: PRO, E101/260/22, 31 Edw. I, 2 February and 20 April 1302.
59. TNA: PRO, E101/261/21, 10–11 Edw. II, 1317.
60. TNA: PRO, E101/260/30, 34–5 Edw. I, 1306.
61. TNA: PRO, E101/260/30, 34 Edw. I, 1306.
62. TNA: PRO, E101/261/25, 16–18 Edw. II, 24 March 1323.
63. TNA: PRO, E101/260/22, 31 Edw. I, 10, 17, 24 August and 7 Sepember etc. 1303.
64. TNA: PRO, E101/263/5, 17 Edw. III; and also TNA: PRO, E 101/263/2, 11–12 Edw. III, 1337–9.
65. TNA: PRO, E101/260/22, 31 Edw. I, 19 October 1303 and 32 Edw. I, 23 November 1303.
66. P. Claughton, *Silver Mining*, appendix 4.
67. *Cal. Fine R.*, vol. 2, p. 126.
68. TNA: PRO, E101/263/5 to 10.
69. TNA: PRO, E101/261/21.
70. L.F. Salzman, 'Mines and stannaries', p. 75.
71. *Ibid*, p. 83.
72. TNA: PRO, E101/263/8.
73. TNA: PRO, E101/260/19.
74. TNA: PRO, E101/260/19.
75. TNA: PRO, E101/261/21.
76. L.F. Salzman, 'Mines and stannaries', p. 83.
77. P. Newman, 'Tinners and tenants on south-west Dartmoor', p. 229.
78. TNA: PRO, E101/260/19.
79. *Cal. Close R., Edw III*, vol. 11, p. 37, 5 June 1360.
80. P. Claughton, *Silver Mining*, appendix 12.
81. T. Slater, 'Medieval town plans', p. 410.
82. *Ibid*, p. 411.

Notes to Chapter 7

1. G. Fairclough and S. Rippon, *Europe's Cultural Landscape*; S. Rippon, *Historic Landscape Analysis*; and see papers in Volume 8.2 of the journal *Landscapes*.

Sources and bibliography

Manuscript sources

Bodleian Library, Oxford
MS Top Devon 62 Material for a history of Devon, J. Milles, mid-eighteenth century

British Geological Survey (BGS), Keyworth, Nottinghamshire
Notebooks of the geologist Henry Dewey

The British Library
Add. MS 24513   Abstracts of Exchequer accounts and miscellaneous records in the Queen's Remembrancers Office
Add. MS 24770   Papers relating to Devonshire
Harl MS 6126 f68   Inquisition post-mortem, Reginald de Ferrers, 34 Edw. I (1305)

Calstock Parish Archive
Transcripts and translations of material relating to the manor and parish of Calstock, Cornwall, carried out under a Manpower Services Commission scheme, drawing on documents from many sources including the Public Record Office and the Duchy Record Office

Cornwall Record Office (CRO), Mount Edgcumbe Papers
ME1739   Survey of the manor of Bere Ferrers, no date (late seventeenth century)
ME1740   Notes for survey of lands at Bere Ferrers, no date (late seventeenth century)
ME2423   Plan of the manor of Bere Ferrers, no date (late seventeenth century)
ME2424   Plan of the manor of Bere Ferrers surveyed for Lord Hobart, 1737
ME2795   Statement of account of dues in Birch and Cleeve mine, 1812–1821

Devon Record Office

1815Z/PZ1–3	Bere Ferrers churchwardens accounts 1600–1656
4672A/HS/79E	Plan of the Old Bere mines

Exeter Dean and Chapter

MS 3522	Manuscript copy of Quinil's summary

Liverpool University: Harold Cohen Library, Special Collections

MS 7.1, 21	Typescript, partial translation of Thomas Cletscher's report to the Swedish Bergskollegium on foreign mines in 1696

The National Archives: Public Record Office (TNA: PRO)

Chancery: Early Proceedings

C1/9/404	Petition of John Clyff, temp. Hen. VI

Chancery: Inquisitions ad quod damnum

C143/192/10	Inquisition into the Combe Martin mines, 1 Edw. III

Exchequer: Accounts

E101/260/4	Account of Vincent de Hulton, 20–22 Edw. I
E101/260/5	Account of Master W de Wymundeham of fine silver received issuing from lead, 20–25 Edw. I
E101/260/6	Account of the same Master W. de Wymondham of fine silver received coming from lead from the king's mines of Birland and of Combe Martin from 27 April 20 Edw. I to 15 Sept. 25 Edw. I
E101/260/7	Accounts of mines in Devon and Cornwall, 21–35 Edw. I
E101/260/17	Documents relating to the sending of miners from Wales, 25 Edw. I
E101/260/19	Account of Thomas de Sweyneseye, keeper of the king's mines in Devonshire and Cornwall, 28–35 Edw. I
E101/260/22	Wage roll, Calstock and Bere Ferrers, 30–32 Edw. I
E101/260/26	Account of wages at the king's mines in Devonshire, 32 Edw. I
E101/260/27	Account of wages at the king's mines, 32–33 Edw. I
E101/260/30	Account of wages at the king's mines, 34 Edw. I
E101/261/10	Counter roll of the Abbot of Tavystock, comptroller, of the revenues of the king's mine in Devon, 26 November, 1 Edw. II to 26 January, 2 Edw. II

E101/261/12	Account of Robert Thorpe, 3–4 Edw. II
E101/261/15	Indenture made between Robertus de Thorp, clerk, keeper of the king's mine in Devon, and Robertus de Berkhamstede, comptroller of the same mine, of the revenue and cash received, Michaelmas 6 Edw. II to the same feast 7 Edw. II
E101/261/21	Account of Ricardus de Wygorn, keeper of the mines 9-10 Edw. II
E101/261/22	Account of John Suge, keeper of the king's mine in Devonshire, 15–16 Edw. II
E101/261/25	Roll of wages and expenses in connection with the mine of Birland, 16–18 Edw. II
E101/262/2	Account of John de Wybringword of expenses of gold mining in Devon and Cornwall, 18 Edw. II
E101/262/4,	Memorandum as to the delivery of the custody of the king's mines in Devonshire by Thomas de Sweyneseye to Richard de Wigorniâ, 18 Edw. II
E101/262/13	Inquisition held at Furshull concerning the mine of Birlande, 1 Edw. III
E101/263/5	Account of John Moveron keeper of the mine of Birland, 16–17 Edw. III
E101/263/7	Counter roll of John Cory, 16–17 Edw. III
E101/263/8	Counter roll of John Cory, mines in Devonshire and Cornwall, 17–18 Edw. III
E101/263/10	Account of John Moveron, 17–18 Edw. III
E101/263/11	Account of John Moveron, keeper of the mine of Birland, 22–23 Edw. III
E101/263/12	Account of mines at Combe Martin, 33–34 Edw. III
E101/265/10 and 11	Particulars of the account of Isabel, late the wife of Richard Curson, 23–29 Hen. VI
E101/265/18	Particulars of the account of James Falleron, 31–36 Hen. VI
E101/266/25	Counter roll of profit and expenses of the silver mine of Bereferrers, 10 July 20 Edw. IV to 10 July 21 Edw. IV
E101/272/19	Account of mining expenses in Devon, 30 Hen. VIII
E101/272/20	Account of mining expenses in Devon, 30 Hen. VIII
E101/273/1	Account of mining expenses in Devon, 30 Hen. VIII

Exchequer: King's Remembrancer: Memoranda Rolls and Enrolment Books

E159/91	Roll for Michaelmas 1317 to Trinity 1318, 11 Edw II

Exchequer: Lord Treasurer's Remembrancer: Memoranda Rolls
E368/73   Roll for Michaelmas 1302 to Trinity 1303, 30/31 Edw. I
E368/85   Roll for Michaelmas 1314 to Trinity 1315, 8 Edw. II

Pipe Rolls
E372/145   Roll for Michaelmas 1299 to Michaelmas 1300
E372/159   Roll for Michaelmas 1313 to Michaelmas 1314
E372/161   Roll for Michaelmas 1315 to Michaelmas 1316
E372/162   Roll for Michaelmas 1316 to Michaelmas 1317

Special Collections: Ancient Correspondence
SC1/48/61   Keeper of the mines to Walter de Langton, treasurer, no date (c.1301)
SC1/48/177   Leadreeve(?) to the Bishop of Bath and Wells, no date

Special Collections: Ancient Petitions
SC8/111/5544   Petition of Reynald de Ferers (Ferrers) of Devon, c.1300
SC8/272/13552   Petition of Thomas de Almaigne, silver finer, c.1324
SC8/322/E550   Petition of Reginald de Ferrers, knight, c.1295-c.1320

Letters and Papers, Henry VIII
SP1/236 f.1   Mines proved by Petrus Filius, c.1528

National Library of Wales (NLW), G.E. Owen Papers
Box 54   Bundle of papers relating to the Combmartin and North Devon Mining Company and shares in the company held by P.G. Jones, solicitor of Carmarthen, 1835–1848

North Devon Athenaeum (NDA), Barnstaple
'Bristol Channel Shipping'   a list of transcripts from coastal port books (E190) held in the National Archives.

West Yorkshire Archive Service (WYAS), Leeds
DB 178/60   Papers relating to shares in the Beeralston and Butspill mines held by Mrs Mary Inchbald, 1809–23

Wiltshire Record Office (WRO)
366/1   Volume of manuscript notes on the Ley family

Printed primary sources

Bracton on the Laws and Customs of England, trans. Samuel E. Thorne (Cambridge, Mass: Belknap Press, 1977)

Calendar of Charter Rolls, 6 vols (London: Public Record Office, 1903–1927)

Calendar of Close Rolls, Henry III, 15 vols (London: Public Record Office, 1902–1975)

Calendar of Close Rolls, Edward I, 5 vols (London: Public Record Office, 1900–1908)

Calendar of Close Rolls, Edward III, 14 vols (London: Public Record Office, 1896–1913)

Calendar of Close Rolls, Richard II, 6 vols (London: Public Record Office, 1914–1927)

Calendar of Documents relating to Scotland, preserved in Her Majesty's Public Record Office, London J. Bain (ed.), 4 vols (Edinburgh: H.M. General Register House, 1881–88)

Calendar of Fine Rolls, 22 vols (London: Public Record Office, 1911–1963)

Calendar of Inquisitions Miscellaneous, 7 vols (London: Public Record Office, 1916–1969)

Calendar of Inquisitions Post Mortem, 16 vols (London: Public Record Office, 1904–1974)

Calendar of Liberate Rolls, 6 vols (London: Public Record Office, 1917–1964)

Calendar of Memoranda Rolls, 1326–1327 (London: Public Record Office, 1969)

Calendar of Patent Rolls, Henry III, 11 vols (London: Public Record Office, 1901–1913)

Calendar of Patent Rolls, Edward I, 4 vols (London: Public Record Office, 1895–1901)

Calendar of Patent Rolls, Henry VI, 6 vols (London: Public Record Office, 1901–1911)

Calendar of Patent Rolls, Edward II, 5 vols (London: Public Record Office, 1894–1904)

Calendar of Patent Rolls, Edward IV (London: Public Record Office, 1897)

Calendar of Patent Rolls, Edward IV – Edward V – Richard III (London: Public Record Office, 1904)

Calendar of Patent Rolls, Edward VI, 6 vols (London: Public Record Office, 1924–1929)

Chronicles of the Reign of Stephen, Henry II and Richard I, ed. R. Howlett, Rolls Series 82, 4 vols (London: Longman, 1884–9)

Domesday Book (gen. ed.) J. Morris (Chichester: Phillimore, 1972–1992)
Letters and Papers, Foreign and Domestic, Henry VIII, 21 vols plus addenda (London: Public Record Office, 1864–1932)
Magnus Rotulus Scaccarii de anno 31 Henrici I, ed. Rev. J Hunter (London, 1833, reproduced in facsimile by HMSO, 1931)
Pipe Rolls, Pipe Roll Society (PRS)
Plowden, E. *The commentaries, or reports of Edmund Plowden: containing divers cases upon matters of law in the several reigns of King Edward VI, Queen Mary, King and Queen Philip and Mary, and Queen Elizabeth* (1588, reprinted London, 1816)
Recuiel de lettres Anglo-Francaises, ed. F.J. Tanquerey (Paris, 1916)
The Mining Journal (London, weekly 1835 on)
The Mining World (London, weekly 1871 on)
The Statutes (London: HMSO, 1950)

Bibliography

Allen, J.R.L., 'A possible medieval trade in iron ores in the Severn estuary of south-west Britain' *Medieval Archaeology* 40 (1996) pp. 226–30.

Allen, J.R.L., 'A medieval pottery assemblage from Magor Pill (Abergwaitha), Caldicot Level – comparative Roman to early-modern trade around the Severn Estuary and beyond' *Archaeology in the Severn Estuary* 14 (2004) pp. 87–110.

Allen, M., 'The volume of the English currency, 1158–1470' *Economic History Review* 2nd ser. 54 (2001) pp. 595–611.

Allen, M., 'Henry II and the English Coinage' in C. Harper-Bill and N. Vincent (eds) Henry II: new interpretations (Woodbridge: Boydell Press, 2007), pp. 257–77.

Agricola, G., *De Re Metallica*, trans. H.C. Hoover and L.H. Hoover (New York: Dover, 1950).

Ancel, B., Cauuet, B. and Cowburn, I., *The Dolaucothi Gold Mines: Archaeological Appraisal* (unpublished report for The National Trust in Wales, 2000).

Anon., *History and Description of the New Bampfylde Copper Mine* (prepared for the British Association meeting at Exeter, 1869).

Ashworth, H.W.W., *Report on Romano-British settlement and metallurgical site* (Wells: Mendip Nature Research Committee Journal Series, March 1970).

Astill, G.G., *A Medieval Industrial Complex and its Landscape: The Metal-working Watermills and Workshops of Bordesley Abbey* (York: Council for British Archaeology, 1993).

Austin, D., Daggett, R.H. and Walker, M.J.C., 'Farms and fields in Okehampton Park, Devon: the problems of studying medieval landscape' *Landscape History* 2 (1980) pp. 39–58.

Austin, D., Gerrard, G.A.M. and Greeves, T.A.P., 'Tin and agriculture in the middle ages and beyond: landscape archaeology in St Neot Parish, Cornwall' *Cornish Archaeology* 28 (1989) pp. 5–251.

Austin, D. and Walker, M.J.C., 'A new landscape context for Houndtor, Devon' *Medieval Archaeology* 29 (1985) pp. 147–51.

Bailly-Maître, M-C. *L'argent, du Minerai au Pouvoir dans la France Médiévale* (Paris: Picard, 2002).

Barnatt, J. and Penny, R., *The Lead Legacy* (Matlock: Peak District National Park Authority in partnership with English Heritage and English Nature, 2004).

Barrow, G.W.S., *The Acts of Malcolm IV, King of Scots, 1153–1165* (Edinburgh: Edinburgh University Press, 1960).

Bartels, C., Fessner, M., Klappauf, L. and Linke, F.A., *Kupfer, Blei und Silber aus dem Goslarer Rammelsberg von den Anfängen bis 1620* (Bocum: Deutches Bergbau-Museum, 2007).

Barton, D.B., *A History of Copper Mining in Cornwall and Devon* (Truro: Truro Bookshop, 1978).

Bayley, J. and Eckstein, K., *Metal-working Debris from Pentrehyling Fort, Brompton, Shropshire* Ancient Monuments Laboratory Report 13/98 (1998).

Beare, T., *The Bailiff of Blackmore* (ed.) J.A. Buckley (Camborne: Penhellick Publications, 1586).

Beddow, A.J.C., *A History of Bere Ferrers Parish*, revised edn (Bere Ferrers: private, 1995).

Beer, K.E. and Scrivener, R.C., 'Metalliferous mineralisation' in E.M. Durrance and D.J.C. Laming (eds) *The Geology of Devon* (Exeter: Exeter University Press, 1982) pp. 117–48.

Beresford, M.W., 'Dispersed and grouped settlement in medieval Cornwall' *Agricultural History Review* 12(i) (1964) pp. 13–27.

Beresford, M.W. and Finberg, H.P.R., *English Medieval Boroughs. A Hand-list* (Newton Abbot: David & Charles, 1973).

Berry, N., *Horner Wood 1995: An Archaeological Survey of Stoke Wood and Ten Acre Cleave* (prepared for the National Trust, held at Holnicote Estate Office, 1995).

Berryman, R.D., *Use of the Woodlands in the Late Anglo-Saxon Period* (Oxford: Archaeopress, 1998).

Bick, D., *Waller's Description of the Mines in Cardiganshire* (Witney: Black Dwarf, 2004).

Birrell, J., 'The peasant craftsman in the medieval forest' *Agricultural History Review* 2nd ser. 17 (1969) pp. 91–107.

Birrell, J., 'The forest and the chase in medieval Staffordshire' *Staffordshire Studies* III (1990–91) pp. 23–50.

Blackburn, M., 'Coinage and currency under Henry I: a review' in M. Chibnall (ed.) *Anglo-Norman Studies XIII* (Woodbridge: Boydell Press, 1991) pp. 49–82.

Blanchard, I.S.W., 'Derbyshire lead production, 1195–1505' *Derbyshire Archaeological Journal* 91 (1971) pp. 119–40.

Blanchard, I.S.W., 'The miner and the agricultural community in late medieval England' *Agricultural History Review* 2nd ser. 20 (1972) pp. 93–106.

Blanchard, I.S.W., 'Rejoinder; Stannator Fabulosus,' *Agricultural History Review* 2nd ser. 22 (1974) pp. 62–74.

Blanchard, I., 'Labour productivity and work psychology in the English mining industry 1400–1600' *Economic History Review* 2nd ser. 31 (1978) pp. 1–24.

Blanchard, I., *International Lead Production and Trade in the "Age of the Saigerprozess" 1460–1560* (Stuttgart: Franz Steiner Verlag, 1995).

Bond, J., 'Medieval charcoal-burning in England' in J. Klapste and P. Sommer (eds) *Arts and Crafts in Medieval Rural Environment* Ruralia VI (Turnhout, Belgium: Brepols Publishers n.v., 2007) pp. 277–94.

Booker, F., *The Industrial Archaeology of the Tamar Valley* (Newton Abbot: David & Charles, 1967).

Boon, G.C., *Cardiganshire Silver and the Aberystwyth Mint in Peace and War* (Cardiff: National Museum of Wales, 1981).

Brand, J.D., 'Some short cross questions' *British Numismatic Journal* 33 (1964), pp. 57–69.

Bray, L.S., *The Archaeology of Iron Production: Romano-British Evidence from the Exmoor Region* (unpublished PhD thesis, University of Exeter, 2007).

Brooke, J., *The Kahlmeter Journal* (Truro: Twelveheads Press, 2001).

Brown, R.A., Colvin, H.M. and Taylor, A.J., *The History of the King's Works, Vol. I: The Middle Ages* (London: HMSO, 1963).

Browne, D. and Hughes, S., *The Archaeology of the Welsh Uplands* (Aberystwyth: Royal Commission on the Ancient and Historical Monuments of Wales, 2003).

Buck, C., *Devon Great Consols: Archaeological Assessment* (unpublished Cornwall Archaeology Unit Report R069, 2002).

Buckley, J.A., *The significance and role of adits in Cornish mine drainage (1700–1900)* (unpublished M.Phil thesis, Camborne School of Mines, 1992).

Burt, R., *The British Lead Mining Industry* (Redruth: Dylansow Truran, 1984).

Burt, R., Waite, P. and Burnley, R., *The Devon and Somerset Mines* (Exeter: Exeter University Press, 1984).
Calvert, J., *The Gold Rocks of Great Britain and Ireland* (London: Chapman and Hall, 1853).
Camm, G.S., *Gold in the Counties of Devon and Cornwall* (St Austell: Cornish Hillside Publications, 1995).
Campbell, B.M.S. and Bartley, K., *England on the Eve of the Black Death* (Manchester: Manchester University Press, 2006).
Campbell, M., 'Gold, silver and precious stones' in J. Blair and N. Ramsay (eds) *English Medieval Industry* (London: The Hambledon Press, 1991) pp. 107–66.
Cannell, J.A., *The Archaeology of Woodland Exploitation in the Greater Exmoor Area in the Historic Period* (Oxford: Archaeopress, 2005).
Cauuet, B., *Les Mines d'Or gauloises du Limousin* (Limoges: Association Culture et Patrimoine en Limousin, 1994).
Cauuet, B., 'Celtic gold mines in west central Gaul' in G. Morteani and J.P. Northover (eds) *Prehistoric Gold in Europe* (Dordrecht: Kluwer, 1995) pp. 219–240.
Cherry, J., 'Leather' in J. Blair and N. Ramsey (eds) *English Medieval Industries: Craftsmen, Techniques and Products* (London: The Hambledon Press, 1991) pp. 295–318.
Cherry, J., 'Pottery and tile' in J. Blair and N. Ramsey (eds) *English Medieval Industries: Craftsmen, Techniques and Products* (London: The Hambledon Press, 1991) pp. 189–209.
Cipolla, C.M., *Before the Industrial Revolution*, 3rd edn (London: Routledge, 1993).
Cirkovic, S., 'The production of gold, silver and copper in the central parts of the Balkans from the thirteenth to the sixteenth century' in H. Kellenbenz (ed.) *Precious Metals in the Age of Expansion* (Stuttgart: Klett-Cotta, 1981) pp. 41–70.
Claughton, P., 'Silver-lead – a restricted resource: technological choice in the Devon mines' in T.D. Ford and L. Willies (eds) *Mining before Powder* (Matlock Bath: Peak District Mines Historical Society / Historical Metallurgy Society, 1994) pp. 54–59.
Claughton, P.F., 'The Lumburn Leat – evidence for new pumping technology at Bere Ferrers in the fifteenth century' in P. Newman (ed.) *The Archaeology of Mining and Metallurgy in South-West Britain* (Matlock Bath: Peak District Mines Historical Society / Historical Metallurgy Society, 1996) pp. 35–40.
Claughton, P., 'Gold at North Molton, and the surviving evidence to be found on the Bampfylde site' *Exmoor Mines Research Group Newsletter* 10 (1997) pp. 2–7; available at http://www.people.exeter.ac.uk/pfclaugh/mhinf/nm_gold.htm.

Claughton, P., *Silver Mining in England and Wales 1066–1500* (unpublished PhD thesis, University of Exeter, 2003).

Claughton, P., 'Production and economic impact: northern Pennine (English) silver in the twelfth century' *Proceedings of the 6th International Mining History Congress*, Akabira City, Japan (2003) pp. 146–49; available for download at http://www.people.exeter.ac.uk/pfclaugh/mhinf/claugh.doc.

Claughton, P., *The Combe Martin Mines*, revised edn (Combe Martin: Combe Martin Local History Group, 2004).

Claughton, P., 'Mining law in England and Wales: understanding boundaries in the landscape' in F. Reduzzi (ed.) *Sfruttamento, tutela e valorizzazione del territorio. Dal diritto romano alla regolamentazione europea e internazionale* (Napoli: Jovene, 2007).

Claughton, P., Paynter, S. and Dunkerley, T (2005) *Further work on lead/silver smelting at Combe Martin, North Devon*, poster presented at the conference on 'Metallurgy – a touchstone for cross-cultural interaction' British Museum, London [PDF document].URL http://www.people.exeter.ac.uk/pfclaugh/mhinf/Lead_si.pdf [accessed on 4 March 2008].

Cleere, H. and Crossley, D., *The Iron Industry of the Weald*, 2nd edn (Cardiff: Merton Priory Press, 1995).

Coates, R., Breeze, A. and Horovitz, D., *Celtic Voices: English Places* (Stamford: Shaun Tyas, 2000).

Craddock, P.T., *Early Metal Mining and Production* (Edinburgh: Edinburgh University Press, 1995).

Craig, Sir J., *The Mint* (Cambridge: Cambridge University Press, 1953).

Crossley, D.W., 'Medieval iron smelting' in D.W. Crossley (ed.) *Medieval Industry* (York: Council for British Archaeology, 1981) pp. 29–41.

Crossley, D., 'White coal and charcoal in the woodlands of north Derbyshire and the Sheffield area', in P. Beswick and I.D. Rotherhamb (eds) *Ancient Woodlands, their archaeology: a coincidence of interest* (Sheffield: Landscape Conservation Forum, 1993), p. 67.

Crump, C.G. and Johnson, C., 'Tables of bullion coined under Edward I, II and III' *The Numismatic Chronicle*, 4th ser., XIII (1913) pp. 200–45.

Dines, H.G., *The Metalliferous Mining Region of South-West England*, 2nd edn (London: HMSO, 1969).

Dixon, D., 'Copper and gold mining in the Exmoor area' in M. Atkinson (ed.) *Exmoor's Industrial Archaeology* (Tiverton: Exmoor Books, 1997) pp. 41–72.

Dunham, K.C., *Geology of the Northern Pennine Orefield, Volume One, Tyne to Stainmore*, 2nd edn (London: HMSO, 1990).

Dunham, K.C., Young, B., Johnson, G.A.L., Colman, T.B., and Fossett, R., 'Rich silver-bearing lead ores in the northern Pennines?' *Proceedings of the Yorkshire Geological Society* 53 (2001) pp. 207–12.

Earl, B., 'Melting tin in the West of England: A study of an old art' *Journal of the Historical Metallurgy Society* 19 (1985) pp. 153–61.

Edmundson, J.C., 'Mining in the later Roman empire and beyond: continuity or disruption' *Journal of Roman Studies* 79 (1989) pp. 84–102.

Ekwall, E., *The Concise Oxford Dictionary of English Place-Names*, 4th edn (Oxford: Oxford University Press, 1960).

Elkington, D., 'The Mendip lead industry' in K. Branigan and P.J. Fowler (eds) *The Roman West Country* (Newton Abbot: David & Charles, 1976) pp. 183–246.

Erskine, A.M., *The Accounts of the Fabric of Exeter Cathedral, 1279–1353, Part 1* Devon and Cornwall Record Society, new series 24 (1981).

Erskine, A.M., *The Accounts of the Fabric of Exeter Cathedral, 1279–1353, Part 2* Devon and Cornwall Record Society, new series 26 (1983).

Everitt, A., 'The marketing of agricultural produce' in J. Thirsk (ed.) *The Agrarian History of England and Wales Volume IV 1500–1640* (Cambridge: Cambridge University Press, 1967) pp. 466–592.

Fairclough, G. and Rippon, S., *Europe's Cultural Landscape* (Brussels: Europae Archaeologiae Consilium, 2002).

Farmer, D.L., 'Woodland and pasture sales on the Winchester manors in the thirteenth century: disposing of a surplus, or producing for the market?' in R.H. Britnell and B.M.S. Campbell (eds) *A Commercialising Economy. England 1086 to c.1300* (Manchester: Manchester University Press, 1995) pp. 102–31.

Field, J., *English Field-Names: A Dictionary* (Newton Abbot: David & Charles, 1982).

Field, J., *A History of English Field-Names* (London: Longman, 1993).

Finberg, H.P.R., 'The stannary of Tavistock' *Report and Transactions of the Devonshire Association for the Advancement of Science, Literature and the Arts* 81 (1949) pp. 155–84.

Finberg, H.P.R., *Tavistock Abbey: A Study in the Social and Economic History of Devon* (Cambridge: Cambridge University Press, 1969).

Finberg, H.P.R., 'The open field in Devon' in H.P.R. Finberg (ed.) *West-Country Historical Studies* (Newton Abbot: David & Charles, 1969) pp. 129–51.

Finberg, H.P.R., 'Morwell' in H.P.R. Finberg (ed.) *West-Country Historical Studies* (Newton Abbot: David & Charles, 1969) pp. 152–68.

Fox, H.S.A., 'Occupation of the land: Devon and Cornwall' in E. Miller (ed.) *The Agrarian History of England and Wales Volume III, 1348–1500* (Cambridge: Cambridge University Press, 1991) pp. 152–74.

Fox, H.S.A., 'Medieval Dartmoor as seen through the account rolls' *Proceedings of the Devon Archaeological Society* 52 (1994) pp. 149–72.

Fox, H.S.A., 'Medieval urban development' in R.J.P. Kain and W. Ravenhill (eds) *Historical Atlas of South-West England* (Exeter: University of Exeter Press, 1999) pp 400–07.

Fyfe, R.M., Brown, A.G. and Rippon, S., 'Mid- to late-Holocene vegetation history of Greater Exmoor, UK: estimating the spatial extent of human-induced vegetation change' *Vegetation History and Archaeobotany* 12 (2003) pp. 215–32.

Fyfe, R.M. and Rippon, S., 'A landscape in transition? Palaeoenvironmental evidence for the end of the 'Romano-British' period in south-west England' in R. Collins and J. Gerrard (eds) *Debating Late Antiquity in Britain* (Oxford: Archaeopress, 2004) pp. 33–42.

Gearey, B.R., West, S. and Charman, D.J., 'The landscape context of medieval settlement in the South-Western moors of England. Recent Palaeoenvironmental evidence from Bodmin Moor and Dartmoor' *Medieval Archaeology* 41 (1997) pp. 195–210.

Gerrard, G.A.M., *The Early Cornish Tin Industry and Archaeological and Historical Survey* (unpublished PhD thesis, University of Wales, 1986).

Gerrard, S., 'Streamworking in medieval Cornwall' *Journal of the Trevithick Society* 14 (1987) pp. 7–31.

Gerrard, S., 'The Dartmoor tin industry: an archaeological perspective' *Proceedings of the Devon Archaeological Society* 52 (1994) pp. 173–98.

Gerrard, S., 'The early south-western tin industry: an archaeological view' in P. Newman (ed.) *The Archaeology of Mining and Metallurgy in South West Britain* (Matlock Bath: Peak District Mines Historical Society / Historical Metallurgy Society, 1996) pp. 67–90.

Gerrard, S., *The Early British Tin Industry* (Stroud: Tempus, 2000).

Gill, M.C., *Lead Mining in the Pennines with Particular Reference to the Duke of Devonshire's Lead Mines at Grassington* (unpublished MPhil. thesis, University of Exeter, 1993).

Gill, M.C., *The Grassington Mines*, British Mining 46 (Keighley: Northern Mine Research Society, 1993).

Gillard, M.J., *The Medieval Landscape of the Exmoor Region: Enclosure and Settlement in an Upland Fringe* (unpublished PhD thesis, University of Exeter, 2002).

Gough, J.W., *Mines of Mendip*, revised edn (Newton Abbot: David & Charles, 1967).

Gover, J.E.B., Mawer, A. and Stenton, F.M., *The Place-Names of Devon, Part I* (Cambridge: Cambridge University Press, 1932).

Greeves, T.A.P., 'The archaeological potential of the Devon tin industry' in D.W. Crossley (ed.) *Medieval Industry* (York: Council for British Archaeology, 1981) pp. 85–95.

Greeves, T.A.P., 'Four Devon stannaries: a comparative study of tin-working in the sixteenth century' in T. Gray, M. Rowe and A. Erskine (eds) *Tudor and Stuart Devon* (Exeter: University of Exeter Press, 1992) pp. 39–74.

Greeves, T.A.P and Rowe, A., 'Four deserted medieval settlements on Buckland Down, West Devon' *Transactions of the Devonshire Association* 131 (1999) pp. 71–80.

Greeves, T.A.P. and Newman, P., 'Tin-working and land-use in the Walkham Valley: a preliminary analysis' *Proceedings of the Devon Archaeological Society* 52 (1994) pp. 199–219.

Griffith, F., *Devon's Past: An Aerial View* (Exeter: Devon County Council, 1988).

Griffith, F. and Weddell, P., 'Ironworking in the Blackdown Hills: results of recent survey' in P. Newman (ed.) *The Archaeology of Mining and Metallurgy in South West Britain* (Matlock Bath: Peak District Mines Historical Society / Historical Metallurgy Society, 1996) pp. 27–34.

Groenewoudt, B., 'Charcoal burning and landscape dynamics in the early medieval Netherlands' in J. Klapste and P. Sommer (eds) *Arts and Crafts in Medieval Rural Environment* Ruralia VI (Turnhout, Belgium: Brepols Publishers n.v., 2007) pp. 327–37.

Gwyn, D., *Gwynedd: Inheriting a Revolution* (Chichester: Phillimore, 2006).

Hall, T.M., 'On the mineral localities of Devonshire' *Report and Transactions of the Devonshire Association for the Advancement of Science, Literature and the Arts* 2 (1868) pp. 332–46.

Hamilton Jenkin, A.K., *Mines of Devon*, new edn (Ashbourne: Landmark Press, 2005).

Hammersley, G., 'Technique or economy; the rise and decline of the early English copper industry' *Business History* 15 (1973) pp. 1–31.

Harris, J.R., *The Copper King* (Liverpool: Liverpool University Press, 1964).

Hart, C., *The Industrial History of Dean: with an introduction to its industrial archaeology* (Newton Abbot: David & Charles, 1971).

Hatcher, J., *Rural Economy and Society in the Duchy of Cornwall 1300–1500* (Cambridge: Cambridge University Press, 1970).

Hatcher, J., *English Tin Production and Trade before 1550* (Oxford: Clarendon Press, 1973).

Hatcher, J., *The History of the British Coal Industry, Volume One, Before 1700: Towards the Age of Coal* (Oxford: Clarendon Press, 1993).

Hatcher, J., 'Myths, miners and agricultural communities' *Agricultural History Review* 22 (1974) pp. 54–61.

Hawken, S., *A Report on the Recent Archaeological Discoveries at Bywood Farm, Dunkeswell, Blackdown Hills, Devon* (unpublished University of Exeter Community Landscapes Project report, 2005).

Hawkins, C., *Vegetation History and Land-use Change in the Blackdown Hills, Devon, UK* (unpublished University of Exeter Community Landscapes Project report, undated).

Hawkes, J.R., 'The Dartmoor granite and later volcanic rocks' in E.M. Durrance and D.J.C. Laming (eds) *The Geology of Devon* (Exeter: University of Exeter Press, 1982) pp. 85–116.

Henderson, C.G. and Weddell, P.J., 'Medieval settlements on Dartmoor and in West Devon: the evidence from excavations', *Proc. Devon Arch. Soc.*, 52 (1994) pp. 119–40.

Henry, A., 'Silver and salvation: a late fifteenth-century confessor's itinerary throughout the parish of Bere Ferrers, Devon, England (Exeter Dean and Chapter MS 3522)' *Report and Transactions of the Devonshire Association for the Advancement of Science, Literature and the Arts* 133 (2001) pp. 17–96.

Henry, A., 'A reply to "Confession in a Fifteenth-Century Parish"' *Report and Transactions of the Devonshire Association for the Advancement of Science, Literature and the Arts* 134 (2002) pp. 69–73.

Herring, P., 'Cornish strip fields' in S. Turner (ed.) *Medieval Devon and Cornwall: Shaping an Ancient Countryside* (Macclesfield: Windgather Press, 2006) pp. 44–77.

Higham, R.A., *The Castles of Medieval Devon* (unpublished PhD thesis, University of Exeter, 1979).

Higham, R.A., 'Castles in Devon' in S. Timms (ed.) *Archaeology of the Devon Landscape: Essays on Devon's Archaeological Heritage* (Exeter: Devon County Council, 1980) pp. 71–80.

Hoare, R.C., *Giraldus Cambrensis, the itinerary through Wales* (London: J.M. Dent, 1935).

Hollister-Short, G., 'On the origins of the suction-lift pump' *History of Technology* 15 (1993) pp. 57–75.

Homer, R.F., 'Tin, lead and pewter' in J. Blair and N. Ramsay (eds), *English Medieval Industry* (London: The Hambledon Press, 1991) pp. 57–80.

Hoskins, W.G., *Devon* (London: Collins, 1954).

Hunter, J.R., 'The medieval glass industry' in D. Crossley (ed.) *Medieval Industry* (London: Council for British Archaeology, 1981) pp. 143–50.

Jaffe, J.A., *The Struggle for Market Power* (Cambridge: Cambridge University Press, 1991).

Jones, A. (ed.), *Flintshire Ministers' Accounts: 1301–1328, extracted from the accounts of the Chamberlains of Chester* (Prestatyn, 1913)

Jones, M., 'South Yorkshire's ancient woodland: the historical evidence' in P. Beswick and I.D. Rotherham (eds) *Ancient Woodlands their archaeology and ecology a coincidence of interest* (Sheffield: Landscape Conservation Forum, 1993) pp. 26–48.

Jones, N., Walters, M. and Frost, P., *Mountains and Orefields: Metal Mining Landscapes of mid- and North-East Wales* (York: Council for British Archaeology, 2004).

Juleff, G., *Early Iron-working on Exmoor* (unpublished report to the Exmoor National Park Authority and the National Trust, Holnicote Estate, 1997).

Juleff, G., 'New radiocarbon dates for iron-working sites on Exmoor' *Historic Metallurgy News* 44 (2000) pp. 3–4.

Juleff, G., 'The Dark Ages remain a mystery' *Exmoor Iron Newsletter* 4 (2006) p. 4.

Kaeuper, R.W., 'The Frescobaldi of Florence and the English Crown' in W.M. Browsky (ed.) *Studies in Medieval and Renaissance History* X (Lincoln: University of Nebraska Press, 1974).

Kirkham, N., *Derbyshire Lead Mining through the Centuries* (Truro: D. Bradford Barton Ltd, 1968).

Knighton, A.D., 'River adjustment to changes in sediment load: the effects of tin mining on the Ringarooma River, Tasmania, 1875–1984' *Earth Surface Processes and Landforms* 14 (1989) pp. 333–59.

Langdon, J. and Masschaele, J., 'Commercial activity and population growth in medieval England' *Past and Present* 190 (February 2006) pp. 35–81.

Leveridge, B.E., Holder, M.T., Goode, A.J.J., Scrivener, R.C., Jones, N.S. and Merriman, R.J., *Geology of the Plymouth and South-East Cornwall Area: Memoir for 1:50,000 Geological Sheet 348* (London: HMSO, 2002).

Lewis, E.A., 'The development of industry and commerce in Wales during the Middle Ages' *Trans. Royal Hist. Soc.*, New Series 17 (1903) pp. 121–74.

Lewis, G.R., *The Stannaries: A study of the English tin mines* (Cambridge, Mass: Harvard University Press, 1908; repr. Truro: D. Bradford Barton Ltd, 1965).

Lewis, P.R. and Jones, G.D.B., 'The Dolaucothi gold mines I: the surface evidence' *The Antiquaries Journal* XLIX (1969) pp. 244–72.

Lysons, D. and Lysons, S., *Magna Britannia: Volume 6, Topographical and Historical Account of Devonshire* (London, 1822).

Macklin, M.G., 'Floodplain sedimentation in the upper Axe valley, Mendip, England' *Transactions of the Institute of British Geographers* 10 (1985) pp. 235–44.

Mackney, D., Hodgson, J.M., Hollis, J.M. and Staines, S.J., *Legend for the 1:250,000 Soil Map of England and Wales* (Soil Survey of England and Wales, 1983).

Maddicott, J.R., 'Trade, industry and the wealth of King Alfred' *Past and Present* 123 (1989) pp. 3–51.

Malham, A., Aylett, J., Higgs, E. and McDonnell, J.G., 'Tin smelting slags from Crift Farm, Cornwall and the effect of changing technology on slag composition' *Historical Metallurgy* 36 (2) (2002) pp. 84–94.

Mayer, P., 'Calstock and the Bere Alston silver-lead mines in the first quarter of the fourteenth century', *Cornish Archaeology* 29 (1990) pp. 79–91.

McDonnell, G., 'Monks and miners: the iron industry of Bilsdale and Rievaulx Abbey' *Medieval Life* 11 (1999) pp. 16–21.

McDonnell, R., *Hawkridge Ridge Wood: An Archaeological Assessment for Management Purposes* (unpublished report to the Exmoor National Park Authority, 1995).

McDonnell, R., *A Preliminary Archaeological Assessment of Eleven Woodlands in the Barle Valley – Hawkridge Area of Exmoor* (unpublished report to the Exmoor National Park Authority, 1999).

Metcalf, D.M., 'The ranking of boroughs: numismatic evidence from the reign of Æthelred II' in A.R. Hands and D.R. Walker (eds) *Ethelred the Unready* (Oxford: Archaeopress, 1978) pp. 159–212.

Mighall, T.M. and Chambers, F.M., 'Early ironworking and its impact on the environment: Palaeoecological evidence from Bryn y Castell Hillfort, Snowdonia, North Wales' *Proceedings of the Prehistoric Society* 63 (1997) pp. 199–219.

Miller, E. and Hatcher, J., *Medieval England: Rural Society and Economic Change 1086–1348* (London: Longman, 1978).

Miller, E. and Hatcher, J., *Medieval England: Towns, Commerce and Crafts, 1086–1348* (London: Longman, 1995).

Moffat, B., 'The environment of Battle Abbey estates (East Sussex) in medieval times; a re-evaluation using analysis of pollen and sediments' *Landscape History* 8 (1986) pp. 77–93.

Moorhouse, S.A., 'Iron production' in M.L. Faull and S.A. Moorhouse (eds) *West Yorkshire: An Archaeological Survey to A.D. 1500, vol. 3* (Wakefield: West Yorkshire Archaeological Service, 1981) pp. 774–79.

Moorhouse, S.A., 'The medieval pottery industry and its markets' in D. Crossley (ed.) *Medieval Industry* (London: Council for British Archaeology, 1981) pp. 96–125.

Myers, A.R., *The Household of Edward IV* (Manchester: Manchester University Press, 1959).
Nayling, N., *The Magor Pill Medieval Wreck* (York: Council for British Archaeology, 1998).
Nef, J.U., 'Mining and metallurgy in medieval civilisation' in M.M. Postan and E. Rich (eds) *The Cambridge Economic history of Europe, Vol. II, Trade and Industry in the Middle Ages* (Cambridge: Cambridge University Press, 1952) pp. 429–567.
Nevell, M., 'Industrialisation, ownership, and the Manchester methodology: the role of the contemporary social structure during industrialisation, 1600–1900' in D. Gwyn and M. Palmer (eds) *Understanding the Workplace, Industrial Archaeology Review*, XXVII: 1 (May 2005) pp. 87–96.
Nevell, M. and Walker, J., *Lands and Lordships in Tameside* (Ashton-under-Lyne: Tameside Metropolitan Borough Council, 1998).
Newman, P., 'The Moorland Meavy – a tinners' landscape' *Report and Transactions of the Devonshire Association for the Advancement of Science, Literature and the Arts* 119 (1987) pp. 223–40.
Newman, P., 'Tinners and tenants on south-west Dartmoor: a case study in landscape history' *Report and Transactions of the Devonshire Association for the Advancement of Science, Literature and the Arts* 126 (1994) pp. 199–238.
Newman, P. (ed.), *The Archaeology of Mining and Metallurgy in South West Britain* (Matlock Bath: Peak District Mines Historical Society / Historical Metallurgy Society, 1996).
Newman, P., 'Tin-working and the landscape of medieval Devon, c.1150–1700' in S. Turner (ed.) *Medieval Devon and Cornwall; Shaping an Ancient Countryside* (Macclesfield: Windgather Press, 2006) pp. 123–43.
Oliver, G., *Monasticon Diocesis Exoniensis* (Exeter: P.A. Hannaford, 1846).
Orme, N., 'Confession in a fifteenth-century Devon parish' *Report and Transactions of the Devonshire Association for the Advancement of Science, Literature and the Arts* 134 (2002) pp. 57–68.
Page, W., *The Victoria County History of Shropshire, Vol. 1* (Oxford: University Press, 1908).
Palmer, M., 'Industrial archaeology' in D. Pearsall (ed.) *Encyclopedia of Archaeology, Vol. 2* (Amsterdam: Elsevier, 2008), pp. 1511–21.
Paynter, S., Dunkerley, T. and Claughton, P., *Lead Smelting Waste from the 2001–2002 Excavations at Combe Martin, Devon* Centre for Archaeology Report 79/2003 (Portsmouth: English Heritage, 2003).
Pennington, R.R., *Stannary Law: A History of the Mining Law of Cornwall and Devon* (Newton Abbot: David & Charles, 1973).

Pevsner, N. and Radcliffe, E., *The Buildings of England: Cornwall*, 2nd edn (Harmondsworth: Penguin, 1970).

Pollard, S. and Crossley, D.W., *The Wealth of Britain, 1085–1966* (London: Batsford, 1968).

Ponting, M., Evans, J.A. and Pashley, V., 'Fingerprinting of Roman mints using laser-ablation MC-ICP-MS lead isotope analysis' *Archaeometry* 45, 4 (2003) pp. 591–97.

Pratt, D., 'Minera: township of the mines' *Trans. Denbighshire Historical Society* 25 (1976) pp. 114–54.

Rackham, O., *Trees and Woodland in the British Landscape*, revised edn (London: Dent, 1993).

Rackham, O., *The History of the Countryside* (London: Phoenix Press, 2000).

Raistrick, A. and Jennings, B., *A History of Lead Mining in the Pennines* (London: Longmans, 1965).

Reed, S.J., *Blackdown Hills Ironworking Project Archaeological Recording of an Iron Ore Extraction Pit, Broadhembury, Devon* (unpublished report no. 97.38, Exeter Archaeology, June 1997).

Rieuwerts, J.H., 'Lead mining in the royal forest of the Peak during the thirteenth century' in T.D. Ford and L. Willies (eds) *Mining before Powder* (Matlock Bath: Peak District Mines Historical Society / Historical Metallurgy Society, 1994) pp. 60–61.

Riley, H. and Wilson-North, R., *The Field Archaeology of Exmoor* (Swindon: English Heritage, 2001).

Rippon, S., *Historic Landscape Analysis* (York: Council for British Archaeology, 2004).

Rippon, S., 'Landscapes of pre-medieval occupation' in R. Kain (ed.) *England's Landscape, Volume 3: The South West* (London: Collins/English Heritage, 2006) pp. 41–66.

Rippon, S., *Landscape, Community and Colonisation* (York: Council for British Archaeology Research Report 152, 2006).

Rippon, S., 'Emerging regional variation in historic landscape character: the possible significance of the "Long Eighth Century"' in M. Gardiner and S. Rippon (eds) *Medieval Landscapes* (Macclesfield: Windgather Press, 2007) pp. 105–21.

Rippon, S., *Beyond the Medieval Village: The Diversification of Landscape Character in Southern Britain* (Oxford: Oxford University Press, in press).

Rippon, S.J., Fyfe, R.M. and Brown, A.G., 'Beyond villages and open fields: the origins and development of a historic landscape characterised by dispersed settlement in South-West England' *Medieval Archaeology* 50 (2006) pp. 31–70.

Roberts, B.K. and Wrathmell, S., *An Atlas of Rural Settlement in England* (London: English Heritage, 2000).

Roberts, C.A. and Gilbertson, D.D., 'The vegetational history of Pile Copse "ancient" oak woodland, Dartmoor, and possible relationships between ancient woodland, clitter and mining' *Proceedings of the Ussher Society* 8 (1994) pp. 298–301.

Rottenbury, J., *Geology, Mineralogy and Mining History of the Metalliferous Mining Area of Exmoor* (unpublished PhD thesis, University of Leeds, 1974).

Rowe, A., *Buckland Monochorum – A West Devon Down and its History* (Privately published, 1999).

Rowley, T., *Villages in the Landscape* (London: Orion Books Ltd, 1994).

Salzman, L.F., 'Mines and stannaries' in J.F. Willard, W.A. Morris and W.H. Dunham (eds) *The English Government at Work 1327–1336, III* (Cambridge, Mass: Medieval Academy of America, 1950) pp. 67–104.

Schubert, H.R., *History of the British Iron and Steel Industry* (London: Routledge & Keagan Paul, 1957).

Searle, E., *Lordship and Community: Battle Abbey and its Banlieu* (Toronto: Pontifical Institute of Mediaeval Studies, 1974).

Shorter, A.H., Ravenhill, W.L.D. and Gregory, K.J., *Southwest England* (London: Nelson, 1969).

Skinner, R.J., 'The "Declaration" of Ley: his pedigree' *Devon and Cornwall Notes and Queries* 37(3) (1993) pp. 101–6.

Slater, T., 'Medieval town plans', in R. Kain and W. Ravenhill (eds) *Historical Atlas of South-West England* (Exeter: University of Exeter Press, 1999) pp. 408-12.

Smart, C. and Claughton, P., 'The mining community and the landscape: the impact of silver mining on the historic landscape in South Devon' in C. Bartels and C. Küpper-Eichas (eds) *Landschaften: Kulturelles Erbe in Europa (Cultural Heritage and Landscapes in Europe)* (Bochum: Deutsches Bergbau-Museum, 2008) pp. 521–43.

Smirke. Sir E., 'Supplementary notes relating to the working of the silver mines in Devon' *Archaeological Journal* 27 (1870) pp. 314–22.

Spufford, P., *Money and its Uses in Medieval Europe* (Cambridge: Cambridge University Press, 1988).

Stos-Gale, S., *Report on the Lead Isotope Analysis of Lead from St David's* (unpublished report for R.J. Lewis, Trefin, Haverfordwest, 2000).

Stuart, E., *Lost Landscapes of Plymouth. Maps, Charts and Plans to 1800* (Stroud: Sutton, 1991).

Substrata Ltd, *Gradiometer and Resistance surveys at Mine Close, Combe Martin* (unpublished report for Mr Trevor Dunkerley, 2004).

Substrata Ltd, *Gradiometer and Resistance surveys at Middleton, Combe Martin* (unpublished report for Mr Trevor Dunkerley, 2004).

Summerson, H., *Crown Pleas of the Devon Eyre of 1238* (Exeter: Devon and Cornwall Record Society, 1985).

Taylor, A.J., *The King's Works in Wales* (London: HMSO, 1974).

Téreygeol, F., 'Les mines d'argent carolingiennes de Melle (Deux-Sèvres): état de la question' *Cahiers Numismatiques* 144 (June 2000) pp. 27–43.

Thorn, C. and Thorn, F (eds), *Domesday Book, 8: Somerset* (Chichester: Phillimore, 1980).

Thorn, C. and Thorn, F (eds), *Domesday Book, 9: Devon* (Chichester: Phillimore, 1985).

Thorndycraft, V.R., Pirrie, D. and Brown, A.G., 'Tracing the record of early alluvial tin mining on Dartmoor, UK' in A.M. Pollard (ed.) *Geoarchaeology: Exploration, Environments, Resources* (London: The Geological Society Publishing House, 1999) pp. 91–102.

Thorndycraft, V.R., Pirrie, D. and Brown, A.G., 'An environmental approach to the archaeology of tin mining on Dartmoor' in P. Murphy and P. Wiltshire (eds) *The Environmental Archaeology of Industry* (Oxford: Oxbow, 2002) pp. 19–28.

Thorndycraft, V.R., Pirrie, D. and Brown, A.G., 'Alluvial records of medieval and prehistoric tin mining on Dartmoor, SW England' *Geoarchaeology* 19(3) (2004) pp. 219–36.

Todd, M., *Roman Mining on Mendip* (Exeter: The Mint Press, 2007).

Tolan-Smith, C., *Landscape Archaeology in Tynedale* (Newcastle upon Tyne: University of Newcastle, 1997).

Toulmin Smith, L (ed.), *The Itinerary of John Leland: In or About the Years 1535–1543. Part I* (London: Bell, 1906–10).

Turner, S., *Ancient Country: The Historic Character of Rural Devon* (Exeter: Devon Archaeological Society Occasional Paper 20, 2007).

Tylecote, R.F., 'Iron Smelting in Pre-Industrial Communities' *Journal of the Iron and Steel Institute*, 203 (1965) pp. 340–48.

Tylecote, R.F., *The Prehistory of Metallurgy in the British Isles* (London: Institute of Metals, 1986).

Wells, W.C., 'The Shrewsbury Mint in the Reign of Richard I, and the Silver Mine at Carreghova' *Numismatic Chronicle* 5th ser. 12 (1932) pp. 215–35.

Westcote, T., *View of Devonshire in MDCXXX* (Exeter: William Roberts, 1845).

Wheeler, J., *The Implications of Iron-Working on the Woodlands of Rievaulx and Bilsdale, North Yorkshire, United Kingdom: Historical, Paleoecological and Paleoenvironmental Perspectives, circa 1068–2000* (unpublished PhD thesis, University of Bradford, 2007).

Wieken, J., *Ironworking in the Blackdown Hills – A report on two slag mounds on Bywood Farm, Dunkeswell and a general discussion of the distribution of iron working* (unpublished University of Exeter Community Landscapes Project report, 2004).

Williams, C.J., 'The mining laws in Flintshire and Denbighshire' in T.D. Ford and L. Willies (eds) *Mining before Powder* (Matlock Bath: Peak District Mines Historical Society, 1994) pp. 62–68.

Williams, R.G.J., 'St Cuthbert's Roman mining settlement, Priddy, Somerset: aerial photographic recognition' *Proceedings of the University of Bristol Speleological Society* 21 (1998) pp. 123–32.

Wilson, J. (ed.), *Victoria History of the County of Cumberland* (London: Archibald Constable and Company Limited, 1901)

Winchester, A., *The Harvest of the Hills: Rural Life in Northern England and the Scottish Borders 1400–1700* (Edinburgh: Edinburgh University Press, 2000).

Zell, M., *Industry in the Countryside: Wealden Society in the Sixteenth Century* (Cambridge: Cambridge University Press, 1994).

Index

(numbers in *italic* are page numbers for figures)

Agricola, Georgius 98–99, *111*, 113
Alston (Cumberland) 48–49

Babewell 67, *68*
Bampfylde Mine, North Molton *14*, 25–26, 40, *41*
Beer Alston 2, 6, 12, *61*, 78, 100, 118, *122*, 126, 128–30, *131–33*, 134–36, 140, 141–45, *146*, 147–49, *152*, 155–5, 158–59, 161, 164
Bere Barton 6, *61*, 136, 138–40
Bere Park 6, *61*, 131–33, 136–38
Beeralston Down 127, *130*, 136, 144, 149
Biccombe (Bicombe, Bickham) 19, *61*, 62, 93–94, 103, 105–6, 153
Birland (Bere Ferrers) 4
Blackdown Hills
  iron working 13, *14*–15, 37–40, 43, 106, 162
Blaxton (Blakstone) *61*, 103, *104*, 105
bole smelting 55–58, 59, 62, 65, 86, 88–92
de Brusele, Henry 157
Brodebirch 84
Brushford *14*, 85
Buckland Abbey *61*, 105, 107, 125, 127, 147

Buckland Monachorum *61*, 62, 89, 90, 100, *135*
bullion crisis 11, 67–69, 110, 163
Buttspill 69, 78, 95, 124, 129, *146*, *152*
Calstock 19, *61*, 89–97, 103, 105, 124–25, *135*, 157, 163
  smelting at, 19, 62, 77, 89, 93–96, 119
Carlisle
  mint 46, *47*
  Mine of 21, 46, 48–50
Carn Galver mine (Cornwall) 2
charcoal 3, 11–12, 28, 39, 42–43, 59, 73, 89, 93, 100, 102, 106–9, 121, 163
Cleave Wood *61*, 75–77, *78*, 80
coal (mineral, sea or stone coal) 108, 120, 179 n. 46
coinage 46–48
de Colle, Richard 157
Columbariis, Phillip de 63
Combe Martin 4, 11, *14*, 15, 18–20, *41*, 54, 54–58, 60, 66, 69, 63–65, 74, 75, 77, 85, 87, 119, 162
confessor's 'itinerary' *152*, 153–55
copper mining 1, 8, 11, 13–14, 16–18, 24–28, 52–53, 66, 69, 110, 162
  at North Molton 24–26, 40, *41*

205

copper smelting 167, n. 64
coppicing 42, 102–4, 120
Crown prerogative 16
cupel / cupellation 22, 45, 71, 97–99
custom / customary regulation 16–17, 19, 21, 23
curia 19, 62, 94–97, 100, 157, 163

Dammewell 67, 68
deadwork / development work 11, 59, 63, 71–74, 83–4, 156
Denham Bridge 105
direct management 17, 19, 54–63, 66, 170, 158, 163
drainage
 by adit 60, 79–84, 110
 pumps 84, 110–19
 suction-lift pumps 111–14
dual occupation 3, 19–20, 34, 49, 54, 59, 83, 86, 155–56, 162
Dulverton *14*, 18, *41*, 42, 85
 cropping units *122*, *147*

Exmoor 13, 15, 24, 27, 28, *41*, 162
 iron working 39–43

family groups, working in 85–86
 de Ferrers, Henry 140
 de Ferrers, John 104, 105, 120
 de Ferrers, Reginald 141, 149, 164
 de Ferrers, William 105, 140
fire-setting 77
Fogge, Sir John 68–69, 113
Frescobaldi 60, 79, 103–4
Frog Street 128, 144
Furzehill
 mine 32, 61, 62, 75, 78, 80–82, 144, *150*, *152*
 settlement *146*

geology 6
geophysical survey *15*, 75, 90, 96

gold working 17, 27–28
 at 'La Hole' 28, 53
Great North Wood 107, 125
Gullytown 67, 75, 125, *146*, *150*, *152*, 175

Hache, William 107
Halsere *61*, 105, 120
Hankelake 107
Higesbale 89–90
High North Shaft 68
Hillgrove 84
historic landscape characterisation (HLC) 121–23, 127, 136, 147–49, *150*, 160
Hochstetter, Joachim 64

iron working 14, 35–44
 in the Blackdown Hills 37–40, 43
 on Exmoor 39–43
 at North Molton 40

Kyrkestey 84

Langleis Shaft *68*
lead production 20–24, 72
 for castle building 21
 from north-east Wales 22–23
lead/silver smelting 88, 119–20
 bellows blown furnace 55–58, 92–94, 97, 108, 119–20
 bole hearth 88–92, *93*
 liquation 89
 turnbole 90
Lockridge Hill 10, *61*, 68, 69, 80, 111–12, 114, 118
Lockridge Pill 144
Lumburn leat 9, *112*, 114–19, 120, 134
Lydford *31*, 32

Maristow (Martinstowe) 61, 62, 78, 92–94, 96, *104*, 105–6, 119, 157, 163, 179 n. 68

Mendip, silver on 3, 45, 46, 48, 50, 85
Metherell 122, 126, 134, 145, 148, 149
mineralisation 6
mineral ownership 16–18, 21–27
miners
    pressed into service 19
    immigrants 158, 162–63
mines at Bere Ferrers: see Furzehill, Maristow, Old, South and South Tamar Consols
Mines Royal, Society of 70
minting 3, 46–48, 50, 52, 56, 59, 60, 66
moor coal (peat charcoal) 33, 41, 73, 108
Morwell Down 5, 6, 9, 124, 149
Morwellham 61, 105, 107, 120

Newhouse 115, 136
North Molton
    iron working 40
    copper mines 24–26

Old Mine 61, 62, 78, 83–84, 112,
open fields 25, 122, 123, 126–34, 140, 143–45, 148, 149, 164
Openhulle, Agnes 86

Park – see Bere Park
Peak District (Derbyshire) 20, 21–23, 70, 86, 156
    miners from 4, 86
Pechis or Peaks Meadow 67, 68
Pennines, north 16
    silver mining in 48–50
prospecting for silver bearing ores 84–85
pumps see drainage

Redgrove 84
refining / refinery see silver

Rodborough (Yodbira) 103, 145

St Dominick 122, 126, 134, 135, 136, 145, 148, 149
silver
    from copper (saigerprocess) 52
    refinery 62, 93–99, 119–20
silver mining
    Swaledale 50
    north-east Wales 50
    at Carreghofa 50
    in continental Europe 50–52
    on Mendip 3, 45, 46, 48, 50, 85
Silver Street 140
Sludde, Richard 86
Smalleye (family) 86
South Mine 61, 62, 74, 78, 83–84, 87
South Tamar Consols 69, 77, 78
Sprynker (or Sprinker), Adrian 67
Styffe Down 68
suction-lift pumps see drainage
Suffolk, Earl of 66–67
Sumpte Shaft 68
Swaneseye, Thomas de 55, 60, 83, 105

Tavistock 5, 30, 61, 63, 102, 104–5, 107, 120, 125, 135, 136, 149
tin smelting 32–33
tin working 13, 14, 15, 22, 28–35
    geoarchaeological evidence 32
Tonnewell 67, 68, 80

Walchemannesknot 83–84
Wales, miners from 4
Warleigh (Warle) 61, 102, 105
Whitsam 61, 62, 67, 75, 122, 126, 131–35, 133, 146, 147, 150–52
woodland 101–20
Wymondham, William de 55–59

Zeal, South 141, 142, 144

www.ingramcontent.com/pod-product-compliance
Lightning Source LLC
Chambersburg PA
CBHW081202240426
43669CB00039B/2781